Electrical Wiring
Commercial

Based on the 1984
NATIONAL ELECTRICAL CODE

Electrical Wiring

Commercial
Fifth Edition

Ray C. Mullin   Robert L. Smith

DELMAR PUBLISHERS INC.

Delmar Staff
    Administrative editor: Mark W. Huth
    Production editor: Ruth Saur

For information, address Delmar Publishers Inc.
2 Computer Drive West, Box 15-015
Albany, New York 12212

Printed in the United States of America
Published simultaneously in Canada
by Nelson Canada,
A division of International Thomson Limited

10 9 8 7 6 5 4 3 2 1

Library of Congress Cataloging in Publication Data

Mullin, Ray C.
    Electrical wiring, commercial.

    "Based on 1984 National Electrical Code."
    Includes index.
    1. Electric wiring, Interior. 2. Mercantile buildings –
– Electric equipment. 3. Electric wiring – Insurance requirements. I.
Smith, Robert L., 1926–   . II. Title.
TK3283.M84   1984   621.31'924   83-21011
ISBN 0-8273-2262-3

CONTENTS

Plans for a Commercial Building (attached to inside back cover):

▬PREFACE▬▬▬▬

The commercial electrician is required to work in three common situations: where the work is planned in advance; where there is no advance planning; and where repairs are needed.

The first situation exists when the work is engineered. In this case, the electrician must know installation procedures and must be able to read plans.

The second situation occurs either during or after construction when changes or remodeling are required.

The third situation arises any time after a system is installed. Whenever a problem occurs with an installation, the electrician must understand the operation of all equipment included in the installation in order to solve the problem.

When the electrician must plan the installation or modify an existing installation, the circuit loads must be determined. *Electrical Wiring — Commercial* is designed to meet these requirements. The text and assignments make frequent reference to the drawings for a commercial building, found in the back packet. Using these drawings develops skill in reading plans. Installation requirements for electrical service, branch circuits, and electrical devices are thoroughly explained. Electrical loads are calculated step-by-step for the commercial building shown on the drawings in the back of the book. All of these topics are covered in conjunction with the *National Electrical Code (NEC)** .

The revised edition of *Electrical Wiring — Commercial* is based on the 1984 *National Electrical Code (NEC)*. The changes in the new Code relating to commercial wiring are thoroughly explained in this text. Illustrations and related information have been updated as necessary, and the commercial building plans (at the back of the text) have been revised to make *Electrical Wiring — Commercial* the most up-to-date guide to wiring commercial buildings.

The student will gain the greatest benefit from the text by using and referring to the *National Electrical Code* on a continuing basis. Certain modifications to the Code rules may be necessary to meet the requirements of state and local electrical codes. The instructor is encouraged to furnish students with any variations from the Code as they affect this commercial installation.

This 1984 edition of *Electrical Wiring — Commercial* has gone through extensive updating for the 1984 *National Electrical Code*. Some of the new material found in this revision are:

• Complete new coverage for recessed fixtures for use in insulated ceilings, in noninsulated ceilings, and in suspended ceilings.

*National Electrical Code® and NEC® are Registered Trademarks of the National Fire Protection Association, Inc., Quincy, MA.

- Simple interpretations for understanding the UL labeling of air conditioning and other similar equipment where the label states "maximum size fuse" or "for use with fuses or HACR breakers."

- Prohibited use of PCB's (polychlorinated biphenyl) in transformers.

- Complete revision on the subject of current carrying capacities of conductors where derating factors and reduction factors must be applied. The subject has been greatly simplified for ease in understanding.

- Some energy-saving considerations by using larger size conductors for a given load. This reduces the heat generated within the conductor.

- Additional coverage of GFCI (ground fault circuit interrupters).

The 1981 edition of the *National Electrical Code* introduced metric (SI) measurements in addition to the traditional English measurements. Accordingly, metric measurements are included in this text, where applicable. Metric conversions are not shown for the dimensions on the commercial plans which accompany this text. Such conversions are considered to be the responsibility of the designer.

This text was prepared by Ray C. Mullin and Robert L. Smith. Mr. Mullin is a former electrical circuit instructor for the Electrical Trades, Wisconsin Schools of Vocational, Technical and Adult Education. A former member of the International Brotherhood of Electrical Workers, Mr. Mullin is presently a member of the International Association of Electrical Inspectors, the Institute of Electrical and Electronic Engineers, Inc., and the National Fire Protection Association, Electrical Section. He completed his apprenticeship training and has worked as a journeyman and supervisor. He has taught both day and night electrical apprentice and journeyman trade extension courses and has conducted engineering seminars. He is currently Regional Manager for a large electrical components manufacturer. Mr. Mullin presents his accumulated knowledge and experience in this text to assist the student in learning commercial wiring in an orderly step-by-step manner. He has been involved in the electrical industry for over 36 years.

Robert L. Smith, P. E., is an Associate Professor of Architecture at the University of Illinois where he teaches courses in electrical systems, lighting design, acoustics, and energy-conscious design. He has worked in electrical construction for more than 20 years, and has taught electrical apprentice classes and courses in industrial electricity.

Professor Smith has been affiliated with a number of professional organizations, including the Illinois Society of Professional Engineers, the International Association of Electrical Inspectors, and the National Academy of Code Administration. He is a lifelong honorary member of the International Brotherhood of Electrical Workers. Professor Smith is presently active with the Illuminating Engineering Society of North America and with committees for Education, Energy Management, and Interior Lighting Applications. He gives frequent public lectures on the application of the *National Electrical Code* and is a regular presentor at the University of Wisconsin Extension.

Electrical Wiring – Commercial, Fifth Edition, has been thoroughly reviewed and tested to ensure its accuracy and usefulness as a learning tool. More than 150,000 students have learned commercial wiring from earlier editions of this text. Each of these students has played a part in making the fifth edition the finest commercial wiring text available. The manuscript for the fifth edition was reviewed by Willard Sexton, Haney Vocational-Technical Center; Jimmie L. Sigmund, Gaston College; and Sidney Graitzer, Lyons Technical Institute.

Electrical Wiring — Commercial is one of Delmar's texts dealing with the electrical trades. Other titles in this category include:

Electrical Wiring — Residential
Electrical Wiring — Industrial
Industrial Electricity
Practical Problems in Mathematics for Electricians
Direct Current Fundamentals
Alternating Current Fundamentals
Electric Motor Control
Electricity 1 — Devices, Circuits, Materials
Electricity 2 — Devices, Circuits, Materials
Electricity 3 — DC Motors and Generators, Controls, Transformers
Electricity 4 — AC Motors, Controls, Alternators
Construction Wiring
Electric Control for Machines
Electricity/Electronics — Principles and Applications

═ACKNOWLEDGMENTS═

American National Standards Institute, Inc., 1430 Broadway, New York, NY 10018.

Appleton Electrical Company, 1747 W. Wellington Avenue, Chicago, IL 60657.

Boltswitch, Inc., 6107 West Lou Avenue, Crystal Lake, IL 60014.

Bussmann Division, McGraw-Edison Company, P.O. Box 14460, St. Louis, MO 63178.

Canadian Standards Association, 178 Rexdale Blvd., Rexdale, Ontario, Canada, M9W 1R3.

Honeywell, Inc., Commercial Div., 2701 Fourth Ave. S., Minneapolis, MN 55408.

Harvey Hubbell, Inc., Wiring Device Div., P.O. Box 3999, Bridgeport, CT 06605.

Illuminating Engineering Society, 345 East 47th Street, New York, NY 10017.

International Association of Electrical Inspectors, 930 Busse Highway, Park Ridge, IL 60068.

Jefferson Electric Co., Div. Litton Industries, 840 South 25th Ave., Bellwood, IL 60104.

Juno Lighting, Inc., 2001 S. Mt. Prospect Road, Des Plaines, IL 60618.

National Electrical Code, National Fire Protection Association, Batterymarch Park, Quincy, MA 02269.

National Electrical Manufacturers Association, 2101 L Street NW, Washington, DC 20037.

Onan Division, Onan Corporation, 1400 73rd Ave. NE, Minneapolis, MN 55432.

Square D Co., Distribution Equip., Group Headquarters, 1601 Mercer Rd., Lexington, KY 40505.

The Trane Co., 3600 Pammel Creek Rd., LaCrosse, WI 54601.

Underwriters Laboratories, Inc., 333 Pfingsten Road, Northbrook, IL 60062.

—UNIT 1—

Commercial Building Plans and Specifications

—OBJECTIVES—

After completing the study of this unit, the student will be able to

- define the job requirements from the contract documents.
- explain the reasons for building plans and specifications.
- locate specific information on the building plans.
- obtain information from industry-related organizations.

COMMERCIAL BUILDING SPECIFICATION

When a building project is awarded, the electrical contractor is given the plans and specifications for the building. These two contract documents govern the construction of the building. It is very important that the contractor and the electricians employed to perform the electrical construction follow the specifications exactly. The electrical contractor will be held responsible for any deviations from the specifications and may be required to correct such deviations or variations at personal expense.

It is suggested that the electrician assigned to a new project first read the specifications carefully. These documents provide the detailed information which will simplify the task of studying the plans. The specifications are usually prepared in book form and may consist of several hundred pages covering all phases of the construction. This text presents in detail only that portion of the specifications which directly affects the electrician; however, summaries of the other specification sections are presented to acquaint the electrician with the full scope of the document.

SPECIFICATION

The specification is a book of rules governing all of the material to be used and the work to be performed on a construction project. The specification is usually divided into several sections.

General Clauses and Conditions

The first section of the specification, titled *General Clauses and Conditions*, deals with the legal requirements of the project. The index to this section may include the following headings:

Notice to Bidders
Schedule of Drawings
Instruction to Bidders
Proposal
Agreement
General Conditions

Some of these items will affect the electrician on the job and others will be of primary concern to the electrical contractor. The following paragraphs give a brief, general description of each item and how it affects either the electrician on the job or the contractor.

Notice to Bidders. This item is of value to the contractor only. The notice describes the project, its location, the time and place of the bid opening, and where and how the plans and specifications can be obtained.

Schedule of Drawings. The schedule is a list, by number and title, of all of the drawings related to the project. Both the contractor and the electrician will use this schedule prior to preparing the bid for the job: the contractor to determine if all the drawings required are at hand, and the electrician to determine if all of the drawings necessary to do the installation are available.

Instructions to Bidders. This section provides the contractor with a brief description of the project, its location, and how the job is to be bid (lump sum, one contract, or separate contracts for the various construction trades, such as plumbing, heating, electrical, and general). In addition, bidders are told where and how the plans and specifications can be obtained prior to the preparation of the bid, how to make out the proposal form, where and when to deliver the proposal, the amount of any bid deposits required, any performance bonds required, and bidders' qualifications. Other specific instructions may be given, depending upon the particular job.

Proposal. The proposal is a form that is filled out by the contractor and submitted at the proper time and place. The proposal is the contractor's bid on a project. The form is the legal instrument that binds the contractor to the owner if: (a) the contractor completes the proposal properly, (b) the contractor does not forfeit the bid bond, (c) the owner accepts the proposal, and (d) the owner signs the agreement. Generally, only the contractor will be using this section.

The proposal may show that alternate bids were requested by the owner. In this case, the electrician on the job should study the proposal and consult with the contractor to learn which of the alternate bids has been accepted in order to determine the extent of the work to be completed.

On occasion, the proposal may include a specified time for the completion of the project. This information is important to the electrician on the job since the work must be scheduled to meet the completion date.

Agreement. The agreement is the legal binding portion of the proposal. The contractor and the owner sign the agreement and the result is a legal contract. Once the agreement is signed, both parties are bound by the terms and conditions given in the specification.

General Conditions. The following items are normally included under the General Conditions heading of the General Clauses and Conditions. A brief description is presented for each item.

- General Note: includes the general conditions as part of the contract documents.

- Definition: as used in the contract documents, this item defines the owner, contractor, architect, engineer, and other people and objects involved in the project.

- Contract Documents: a listing of the documents involved in the contract, including plans, specifications, and agreement.

- Insurance: specifies the insurance a contractor must carry on all employees, and on the materials involved in the project.

- Workmanship and materials: specifies that the work must be done by skilled workers and that the materials must be new and of good quality.

- Substitutions: materials used must be as specified or equivalent materials must be shown to have the required properties.

- Shop drawings: this item identifies the drawings which must be submitted by the contractor to show how the specific pieces of equipment are to be installed.

- Payments: the method of paying the contractor during the construction is specified.

- Coordination of Work: each contractor on the job must cooperate with every other contractor to insure that the final product is complete and functional.

- Correction Work: describes how work must be corrected, at no cost to the owner, if any part of the job is installed improperly by the contractor.

- Guarantee: the contractor guarantees the work for a certain length of time, usually one year.

- Compliance with all laws and regulations: specifies that the contractor will perform all work in accordance with all required laws, ordinances, and codes such as the *National Electrical Code* and city codes.

- Others: these sections are added as necessary by the owner, architect, and engineer when the complexity of the job and other circumstances require them. None of the items listed in the General Conditions has precedence over another item in terms of its effect on the contractor or the electrician on the job. The electrician must study each of the items before taking a position and assuming responsibilities with respect to the job.

SUPPLEMENTARY GENERAL CONDITIONS

The second main section of the specifications is titled *Supplementary General Conditions*. These conditions usually are more specific than the general conditions. While the general conditions can be applied to any job or project in almost any location with little change, the supplementary general conditions are rewritten for each project. The following list covers the items normally specified by the supplementary general conditions.

- The contractor must instruct all crews to exercise caution while digging; any utilities damaged during the digging must be replaced by the contractor responsible.

- The contractor must verify the existing conditions and measurements.

- The contractor must employ qualified individuals to lay out the work site accurately. A registered land surveyor or engineer may be part of the crew responsible for the layout work.

- Job offices are to be maintained as specified on the site by the contractor; this office space may include space for owner representatives.

- The contractor may be required to provide telephones at the project site for use by the architect, engineer or owner.

- Temporary toilet facilities and water are to be provided by the contractor for the construction personnel.

- Temporary light and power. The contractor must supply an electrical service of a specified size to provide temporary light and power at the site.

- It may be necessary for the contractor to supply a specified type of temporary heating to keep the temperature at the level specified for the structure.

- According to the terms of the guarantee, the contractor agrees to replace faulty equipment and correct construction errors for a period of one year.

The above listing is by no means a complete catalog of all of the items that can be included in the section on *Supplementary General Conditions*.

Other names may be applied to the *Supplementary General Conditions* section; these names include *Special Conditions* and *Special Requirements*. Regardless of the name used, these sections contain the same types of information. All sections of the specifications must be read and studied by all of the construction trades involved. In other words, the electrician must study the heating, plumbing, ventilating, air conditioning, and general construction specifications to determine if there is any equipment furnished by the other trades, where the contract specifies that such equipment is to be installed by the electrical contractor. The electrician must also study the general construction specifications since the roughing in of the electrical system will depend on the types of construction that will be encountered in the building.

This overview of the general and supplementary general conditions of a specification is intended to show the student that the construction worker on the job is affected by parts of the specification other than the part designated for each particular trade.

Contractor Specification

In addition to the sections of the specification which apply to all contractors, there are separate sections for each of the contractors, such as the general contractor who constructs the building proper, the plumbing contractor who installs the water and sewage systems, the heating and air-conditioning contractor, and the electrical contractor. The contract documents usually do not make one contractor responsible for work specified in another section of the specifications. However, it is always considered good practice for each contractor to be aware of being involved in each of the other contracts in the total job.

COMMERCIAL BUILDING PLANS

The construction plans for a building are usually called *blueprints*. This term is a carry-over from the days when the plans were blue with white lines. Today, a majority of the plans used have black lines on white since this combination is considered easier to read.

A set of 13 plan sheets is included at the back of the text showing the general and electrical portions of the work specified.

- *Sheet 1 – Plot Plan:* shows the location of the building and may give grading elevations.

- *Sheets 2, 3, and 4 – Floor Plans:* give the wall and partition details for the building. These sheets are dimensioned; the electrician can find exact locations by referring to these sheets. The electrician should also check the plans for the materials used in the general construction as these will affect when and how the system will be installed.

- *Sheets 5, 6, 7, and 8 – Elevations and Details:* the four exterior views of the building are shown. The electrician must study the elevation dimensions which are given in hundredths of a foot above sea level. For example, the finished second floor, which is shown at 218.33', is 218 feet 4 inches above sea level. These sheets also include *detail* drawings. Since it may be necessary for the electrician to refer to these details, their location should be known.

- *Sheets 9 and 10 – Sections and Details:* these sheets give detail drawings of the more important sections of the building. The location of the section is indicated on the floor plans. When looking at a section, imagine that you are looking in the direction of the arrows at a building that is cut in two at the place indicated. You should see the section exactly as you view the imaginary building from this point.

- *Sheets E1, E2, and E3 – Electrical:* these sheets show the detailed electrical work on an outline of the building. Since dimensions usually are not shown on the electrical plans, the electrician must consult the other sheets for this information. It is recommended that the electrician refer frequently to the other plan sheets to insure that the electrical installation does not conflict with the work of the other construction trades.

To assist the electrician in recognizing components used by other construction trades, the following illustrations are included: figure 1-1, Architectural Drafting Symbols; figure 1-2, Standard Symbols for Plumbing, Piping and Valves; figure 1-3, Sheet Metal Ductwork Symbols; and figure 1-4, Electrical Symbols for Architectural Plans. However, the electrician should be aware that variations of these symbols may be used and the specification and/or plans for a specific project should always be consulted.

CODES AND ORGANIZATIONS

Local Codes

Many organizations such as cities and power companies develop electrical codes which they enforce within their areas of influence. These codes generally are concerned with the design and installation of electrical systems. In many cases, the *National Electrical Code* is used as the basis for the local code. It is always advisable to consult these organizations before work is started on any project. The local codes may contain special requirements which apply to the installation. Additionally, the contractor may be required to obtain special permits and/or licenses before construction work can begin.

National Fire Protection Association

Organized in 1896, the National Fire Protection Association (NFPA) is an international, non-profit organization dedicated to the twin goals of promoting the science of fire protection and improving fire protection methods. The NFPA annually publishes a 10-volume series covering the national fire codes. The *National Electrical Code (NEC)* is a part of volume 5 of this series. The purpose and scope of the *NEC®** are discussed in *Article 90* of the Code.

*National Electrical Code® and NEC® are Registered Trademarks of the National Fire Protection Association, Inc., Quincy, MA.

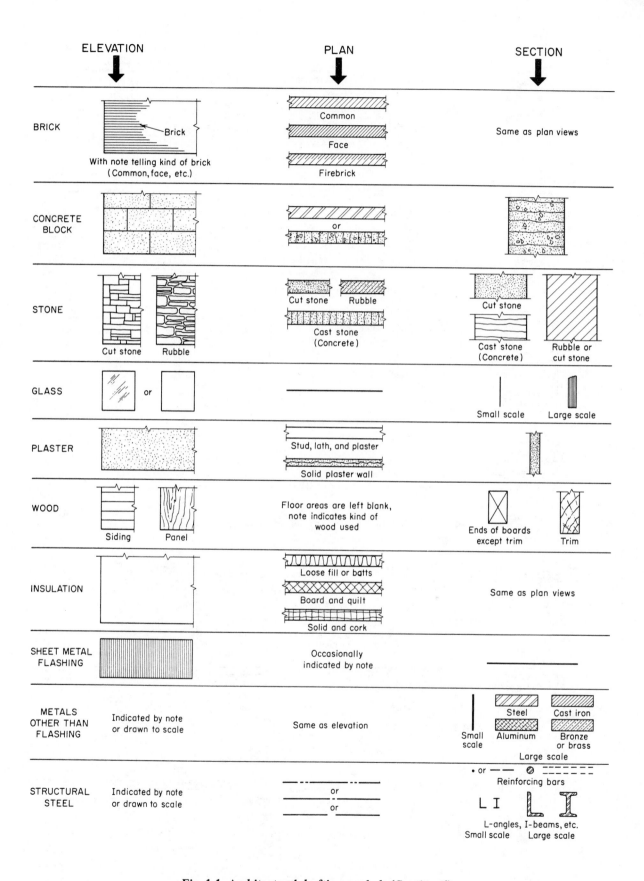

Fig. 1-1 Architectural drafting symbols (Continued).

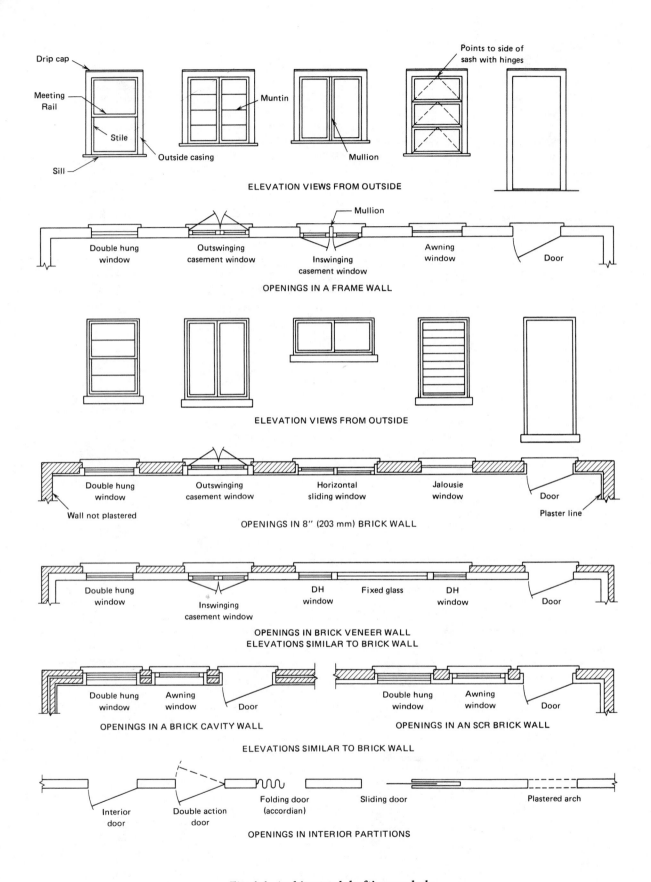

Fig. 1-1 Architectural drafting symbols.

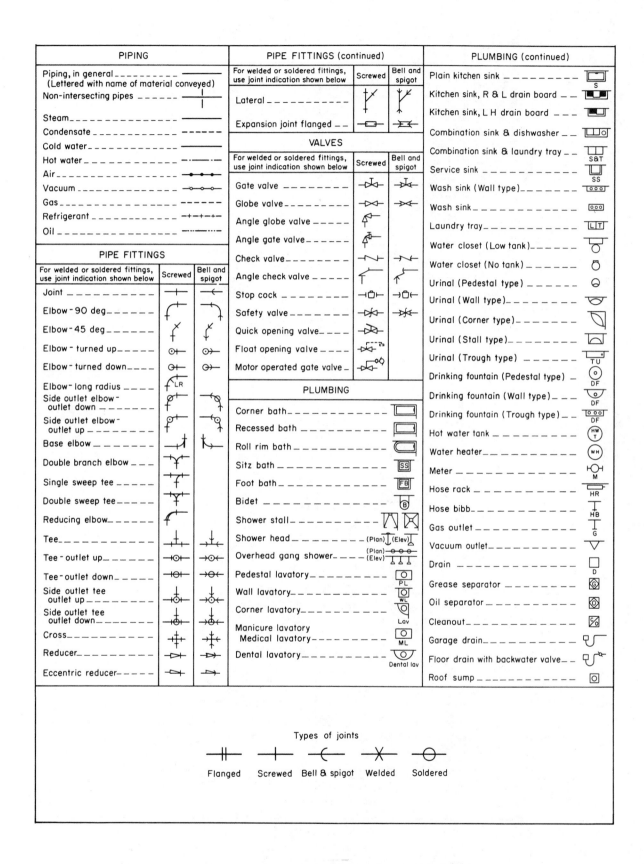

Fig. 1-2 Standard symbols for plumbing, piping and valves.

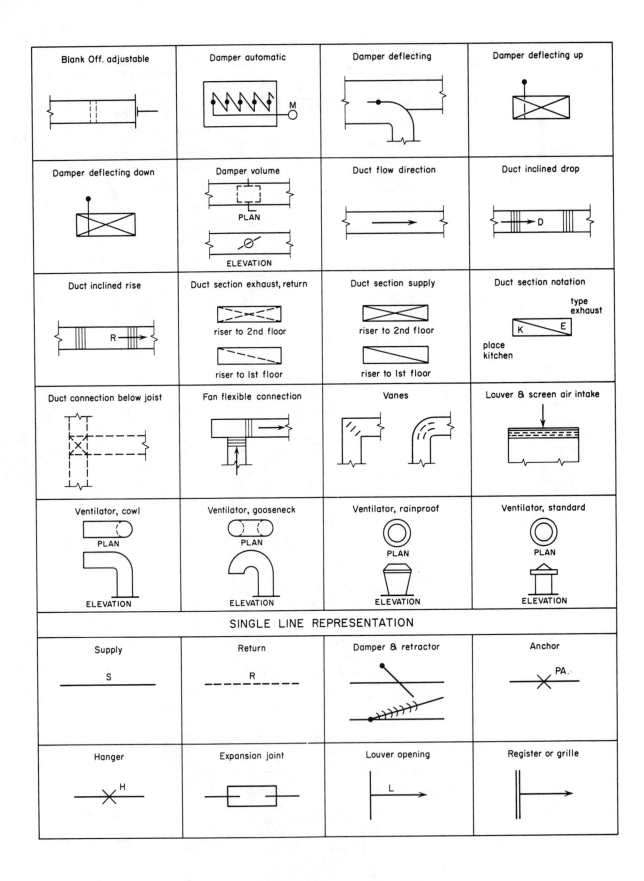

Fig. 1-3 Sheet metal ductwork symbols.

GENERAL OUTLETS

CEILING	WALL	
		Outlet.
Ⓑ	Ⓑ	Blanked Outlet.
Ⓓ	Ⓓ	Drop Cord.
Ⓔ	Ⓔ	Electrical Outlet—for use only when circle used alone might be confused with columns, plumbing symbols, etc.
Ⓕ	Ⓕ	Fan Outlet.
Ⓙ	Ⓙ	Junction Box.
Ⓛ	Ⓛ	Lampholder.
Ⓛ PS	Ⓛ PS	Lampholder with Pull Switch.
Ⓢ	Ⓢ	Pull Switch.
Ⓥ	Ⓥ	Outlet for Vapor Discharge Lamp.
Ⓧ	Ⓧ	Exit Light Outlet.
Ⓒ	Ⓒ	Clock Outlet. (Specify Voltage).

RECEPTACLE OUTLETS

Single Receptacle Outlet.

Duplex Receptacle Outlet.

Triple Receptacle Outlet.

Duplex Receptacle Outlet, Split Circuit.

Duplex Receptacle Outlet with NEMA 5-20R Receptacle.

Weatherproof Receptacle Outlet.

Range Receptacle Outlet.

Switch and Receptacle Outlet.

Radio and Receptacle Outlet.

Special Purpose Receptacle Outlet.

Floor Receptacle Outlet.

SWITCH OUTLETS

S	Single-pole Switch.
S2	Double-pole Switch.
S3	Three-way Switch.
S4	Four-way Switch.
SD	Automatic Door Switch.
SE	Electrolier Switch.
SK	Key Operated Switch.
SP	Switch and Pilot Lamp.
SCB	Circuit Breaker.
SWCB	Weatherproof Circuit Breaker.
SMC	Momentary Contact Switch.
SRC	Remote Control Switch.
SWP	Weatherproof Switch.
SF	Fused Switch.
SWF	Weatherproof Fused Switch.

SPECIAL OUTLETS

O a,b,c- etc.

⊖ a,b,c- etc.

S a,b,c- etc.

Any Standard Symbol as given above with the addition of a lower case subscript letter may be used to designate some special variation of Standard Equipment of particular interest in a specific set of architectural plans.

When used they must be listed in the Key of Symbols on each drawing and if necessary further described in the specifications.

PANELS, CIRCUITS AND MISCELLANEOUS

Lighting Panel.

Power Panel.

Branch Circuit; Concealed in Ceiling or Wall.

Branch Circuit; Concealed in Floor.

Branch Circuit; Exposed.

Home Run to Panelboard. Indicate number of Circuits by number of arrows.

Note: Any circuit without further designation indicates a two-wire circuit. For a greater number of wires indicate as follows:

/// (3 wires) //// (4 wires), etc.

Feeders. Note: Use heavy lines and designate by number corresponding to listing in Feeder Schedule.

Underfloor Duct and Junction Box. Triple System. Note: For double or single systems eliminate one or two lines. This symbol is equally adaptable to auxiliary system layouts.

Ⓖ Generator.

Ⓜ Motor.

Ⓘ Instrument.

Ⓣ Power Transformer. (Or draw to scale.)

⊠ Controller.

Isolating Switch.

Overcurrent device, (fuse, breaker, thermal overload)

Switch and fuse

AUXILIARY SYSTEMS

Push Button.

Buzzer.

Bell.

Annunciator.

Outside Telephone.

Interconnecting Telephone.

Telephone Switchboard.

Bell Ringing Transformer.

D Electric Door Opener.

F Fire Alarm Bell.

F Fire Alarm Station.

X City Fire Alarm Station.

FA Fire Alarm Central Station.

FS Automatic Fire Alarm Device.

W Watchman's Station.

W Watchman's Central Station.

H Horn.

N Nurse's Signal Plug.

M Maid's Signal Plug.

R Radio Outlet.

SC Signal Central Station.

Interconnection Box.

Battery.

Auxiliary System Circuits.

Note: Any line without further designation indicates a 2-wire System. For a greater number of wires designate with numerals in manner similar to 12—No. 18W-3/4''-C., or designated by number corresponding to listing in Schedule.

Special Auxiliary Outlets.

a,b,c Subscript letters refer to notes on plans or detailed description in specifications.

Fig. 1-4 Electrical symbols for architectural plans.

Although the NFPA is an advisory organization, the recommended practices contained in its published codes are widely used as a basis for local codes. Additional information concerning the publications of the NFPA and membership in the organization can be obtained by writing to:

National Fire Protection Association, Inc.
Batterymarch Park
Quincy, Massachusetts 02269

National Electrical Code

The original *National Electrical Code* was developed in 1897. Sponsorship of the Code was assumed by the NFPA in 1911. The purpose of the *National Electrical Code (NEC)* is stated in *Article 90,* Introduction, *Section 90-1.*

The *National Electrical Code* generally is the "bible" for the electrician. However, the *NEC* did not have a legal status until the appropriate authorities adopted it as a legal standard. In May 1971, the Department of Labor, through the Occupational Safety and Health Administration (OSHA), adopted the *NEC* as a national consensus standard. Therefore, in the areas where OSHA is enforced, the *NEC* is law.

Throughout this text, references are made to articles and sections of the *National Electrical Code.* It is suggested that the student, and any person interested in electrical construction, obtain and use a copy of the latest edition of the *NEC.* To help the user of this text, relevant Code sections are paraphrased where appropriate. However, the *NEC* must be consulted before any decision related to electrical installation is made.

Code Terms. The following terms are used throughout the Code. It is important to understand the meanings of these terms.

APPROVED: Acceptable to the authority having jurisdiction.

AUTHORITY HAVING JURISDICTION: An organization, office, or individual responsible for "approving" equipment, an installation, or a procedure.

IDENTIFIED: (As applied to equipment.) Recognizable as suitable for a specific purpose, function, use, environment, or application, where described in a particular Code section. For example, "Identified for use in a wet location."

LABELED: Equipment or materials to which has been attached a label, symbol, or other identifying mark of an organization acceptable to the authority having jurisdiction and concerned with product evaluation, that maintains periodic inspection of production of labeled equipment or materials and by whose labeling the manufacturer indicates compliance with appropriate standards or performance in a specified manner.

LISTED: Equipment or materials included in a list published by an organization acceptable to the authority having jurisdiction and concerned with product evaluation, that maintains periodic inspection of production of listed equipment or materials and whose listing states either that the equipment or material meets appropriate standards or has been tested and found suitable for use in a specified manner.

SHALL: Indicates a mandatory requirement.

SHOULD: Indicates a recommendation or that which is advised but not required.

Copies of the *NEC* are available from most bookstores or from:

National Fire Protection Association, Inc.
Batterymarch Park
Quincy, Massachusetts 02269

Underwriters Laboratories, Inc.

Founded in 1894, Underwriters Laboratories (UL) is a nonprofit organization which operates laboratories to investigate materials, devices, products, equipment, and construction methods and systems to define any hazards which may affect life and property. The organization provides a listing service to manufacturers, figure 1-5. Any product authorized to carry an Underwriters' listing has been evaluated with respect to all reasonable forseeable hazards to life and property, and it has been determined that the product provides safeguards to these hazards to an acceptable degree. A listing by the Underwriters Laboratories does not mean that a product is *approved* by the *National Electrical Code.* However, many local agencies do make this distinction.

The *National Electrical Code* defines *approved* as ". . .acceptable to the authority having jurisdiction. . ." *Section 110-3(b)* of the *NEC* requires that all electrical equipment that is listed and labeled must be used or installed in accordance with any instruction that may be included in the listing or labeling.

An index of publications and information concerning the Underwriters Laboratories can be obtained by writing to:

Public Information
Underwriters Laboratories, Inc.
333 Pfingsten Road
Northbrook, Illinois 60062

National Electrical Manufacturers Association

The National Electrical Manufacturers Association (NEMA) is a nonprofit organization supported by the manufacterers of electrical equipment and supplies. NEMA develops standards which are designed to assist the purchaser in selecting and obtaining the correct product for specific applications, figure 1-6. Information concerning NEMA standards may be obtained by writing to:

National Electrical Manufacturers Association
2101 L Street, N.W.
Washington, DC 20037

American National Standards Institute, Inc.

The American National Standards Institute, Inc. (ANSI) is located at 1430 Broadway, New York, New York 10018. Various working groups in the organization study the numerous codes and standards. An American National Standard implies "a consensus of those concerned with its scope and provisions." The *National Electrical Code* is approved by ANSI and is numbered ANSI/NFPA 70-1984.

Canadian Standards Association

The Canadian Electrical Code is similar to the *National Electrical Code*. Those using this text in

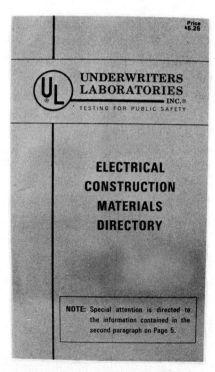

Fig. 1-5 Underwriters Laboratories, Inc.

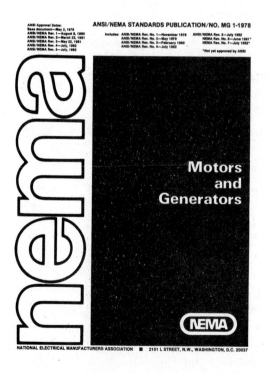

Fig. 1-6 National Electrical Manufacturers Association. (This material is reproduced by permission of the National Electrical Manufacturers Association from NEMA Standards Publication No. MG 1-1978, *Motors and Generators*, copyright 1982 by NEMA)

Canada must follow the Canadian Electrical Code. The Code, which was in effect at the time of preparing this text, is known as the "Safety Standards for Electrical Installations," Standard C22.1.

The Canadian Code is a voluntary code suitable for adoption and enforcement by electrical inspection authorities. The Canadian Electrical Code is available from:

Canadian Standards Association
178 Rexdale Blvd.
Rexdale, Ontario, Canada
M9W 1R3

International Association of Electrical Inspectors

The International Association of Electrical Inspectors (IAEI) is a nonprofit organization. The IAEI membership consists of electrical inspectors, electricians, contractors, and manufacturers throughout the United States and Canada. One goal of the IAEI is to improve the understanding of the *NEC*. Representatives of this organization serve as members of the various panels of the *National Electrical Code* Committee and share equally with other members in the task of reviewing and revising the *NEC*. The IAEI publishes a bimonthly magazine, the *IAEI News*. Additional information concerning the organization may be obtained by writing to:

International Association of Electrical Inspectors
930 Busse Highway
Park Ridge, Illinois 60068

Illuminating Engineering Society of North America

The Illuminating Engineering Society of North America (IESNA) was formed more than 65 years ago. The objective of this group is to communicate information about all facets of good lighting practice to its members and to consumers. The IESNA produces numerous publications which are concerned with illumination.
The *IESNA Lighting Handbooks* are regarded as the standard for the illumination industry and contain essential information about light, lighting, and luminaires. Information about publications or membership may be obtained by writing to:

Illuminating Engineering Society
of North America
345 East 47th Street
New York, New York 10017

Registered Professional Engineer

Although the requirements may vary slightly from state to state, the general statement can be made that a registered professional engineer has demonstrated his or her competence by graduating from college and passing a difficult licensing examination. Following the successful completion of the examination, the engineer is authorized to practice engineering under the laws of the state. A requirement is usually made that a registered professional engineer must supervise the design of any building that is to be used by the public. The engineer must indicate approval of the design by affixing a seal to the plans.

Information concerning the procedure for becoming a registered professional engineer and a definition of the duties of the professional engineer can be obtained by writing the state government department which supervises licensing and registration.

NEC USE OF METRIC (SI) MEASUREMENTS

The *National Electrical Code* includes both English and metric measurements. The metric system is known as the *International System of Units* (SI).

Metric (SI) measurements appear in the Code as follows:

- in the Code paragraphs, the approximate metric (SI) measurement appears in parentheses following the English measurement.

- in the Code tables, a footnote shows the SI conversion factors.

A metric (SI) measurement is not shown for conduit size, box size, wire size, horsepower designation for motors, and other "trade sizes" that do not reflect actual measurements.

Guide to Metric (SI) Usage

The SI metric system is the international system of measurement. In many countries commas are used as decimal markers. Therefore, in the metric system, commas are not used to separate digits. Digits are grouped into sets of three counting both left and right of the decimal point with a space used to separate the groups (e.g., 1 000.000 26). However, this text uses the English method of separating digits, the comma.

In the metric (SI) system, the units increase or decrease in multiples of 10, 100, 1 000, and so on. For instance, one megawatt (1 000 000 watts) is 1 000 times greater than one kilowatt (1 000 watts).

By assigning a name to a measurement, such as a *watt*, the name becomes the unit. Adding a prefix to the unit, such as *kilo*, forms the new name *kilowatt*, meaning 1 000 watts. Refer to figure 1-7 for prefixes used in the metric (SI) system.

Certain of the prefixes shown in figure 1-8 have a preference in usage. These prefixes are *mega, kilo*, the unit itself, *centi, milli, micro,* and *nano.* Consider that the basic unit is a meter (one). Therefore, a kilometer is 1 000 meters, a centimeter is 0.01 meter, and a millimeter is 0.001 meter. Thus, 900 millimeters can be expressed as 0.9 meter.

The advantage of the metric (SI) system is that there is less possibility of confusion by recognizing the meaning of the proper prefix. For example, a four-foot lamp is approximately 1 200 millimeters, or 1.2 meters.

Some common measurements of length in the English system are shown with their metric (SI) equivalents in figure 1-8.

Electricians will find it useful to refer to the conversion factors and their abbreviations shown in figure 1-9.

mega	1 000 000	(one million)
kilo	1 000	(one thousand)
hecto	100	(one hundred)
deka	10	(ten)
the unit	1	(one)
deci	0.1	(one-tenth) (1/10)
centi	0.01	(one-hundredth) (1/100)
milli	0.001	(one-thousandth) (1/1 000)
micro	0.000 001	(one-millionth) (1/1 000 000)
nano	0.000 000 001	(one-billionth) (1/1 000 000 000)

Fig. 1-7 Metric (SI) prefixes and their values.

one inch	=	2.54	centimeters
	=	25.4	millimeters
	=	0.025 4	meter
one foot	=	12	inches
	=	0.304 8	meter
	=	30.48	centimeters
	=	304.8	millimeters
one yard	=	3	feet
	=	36	inches
	=	0.914 4	meter
	=	914.4	millimeters
one meter	=	100	centimeters
	=	1 000	millimeters
	=	1.093	yards
	=	3.281	feet
	=	39.370	inches

Fig. 1-8 Some common measurements of length and their metric (SI) equivalents.

inches (in) $\times$ 0.025 4 = meter (m)
inches (in) $\times$ 0.254 = decimeters (dm)
inches (in) $\times$ 2.54 = centimeters (cm)
centimeters (cm) $\times$ 0.393 7 = inches (in)
inches (in) $\times$ 25.4 = millimeters (mm)
millimeters (mm) $\times$ 0.039 37 = inches (in)
feet (ft) $\times$ 0.304 8 = meters (m)
meters (m) $\times$ 3.280 8 = feet (ft)
square inches (in^2) $\times$ 6.452 = square centimeters (cm^2)
square centimeters (cm^2) $\times$ 0.155 = square inches (in^2)
square feet (ft^2) $\times$ 0.093 = square meters (m^2)
square meters (m^2) $\times$ 10.764 = square feet (ft^2)
square yards (yd^2) $\times$ 0.836 1 = square meters (m^2)
square meters (m^2) $\times$ 1.196 = square yards (yd^2)
kilometers (km) $\times$ 1 000 = meters (m)
kilometers (km) $\times$ 0.621 = miles (mi)
miles (mi) $\times$ 1.609 = kilometers (km)

Fig. 1-9 Useful conversions (English/SI-SI/English) and their abbreviations.

REVIEW

Note: Refer to the *National Electrical Code* or the plans where necessary.

1. The requirements for temporary light and power at the job site will be found in what portion of the specifications? _____

2. The electrician uses the Schedule of Drawings for what purpose?

   _____

3. What section of the specification contains a list of the contract documents?

   _____

Complete items 4-13 by inserting the letter of the correct source of information for:

4. _____ Room width
5. _____ Grading elevations
6. _____ Ceiling height
7. _____ Floor construction
8. _____ Window head construction
9. _____ Exterior wall finishes
10. _____ View of interior wall
11. _____ Outlet locations
12. _____ Receptacle style
13. _____ Swing of a door

A. Plot plan
B. Architectural floor plan
C. Elevations
D. Sections
E. Details
F. Electrical floor plans
G. Specification

Match the following items:

14. _____ NFPA
15. _____ *NEC*
16. _____ Underwriters Laboratories, Inc.
17. _____ NEMA
18. _____ IAEI
19. _____ Registered Professional Engineer
20. _____ IESNA

A. Lighting handbooks
B. Seal
C. Manufacturers' standards
D. 10-volume fire codes
E. Listing service
F. Electrical inspectors
G. Electrical Code

UNIT 2

The Electric Service

OBJECTIVES

After completing the study of this unit, the student will be able to

- install power transformers to meet the *NEC* requirements.
- draw the basic transformer connection diagrams.
- recognize different service types.
- connect metering equipment.
- apply ground-fault requirements to an installation.
- install a grounding system.

The installation of the electric service to a building requires the cooperation of the electrician and the local power company. The availability of high voltage and the power company requirements determine the type of service to be installed. This unit will investigate several common variations in electrical service installations together with the applicable *NEC* regulations.

ASKAREL-INSULATED AND OIL-INSULATED TRANSFORMERS

Many transformers are immersed in a liquid which may be askarel or oil. This liquid performs several important functions: (1) it is part of the required insulation, and (2) it acts as a coolant by conducting heat from the core and the winding of the transformer to the surface of the enclosing tank which then is cooled by radiation. Askarel is a nonflammable synthetic liquid having excellent electrical insulating qualities. Such transformers need not be installed in a vault unless the primary voltage is more than 35,000 volts. Askarel-insulated transformers are used for indoor installations rather than oil-filled transformers. The installation

of askarel-insulated and oil-insulated transformers is covered in *Sections 450-23, 450-24, 450-25,* and *450-26 of the NEC.*

Askarel is classified as a PCB (polychlorinated biphenyl) liquid. The Environmental Protection Agency has set forth severe restrictions on the use of PCB liquids as the liquid filler for transformers, capacitors, and similar equipment. Manufacturers of transformers now use a nonPCB liquid. However, the *National Electrical Code* presently does not involve itself in environmental topics and continues to recognize askarel as an excellent insulating liquid.

DRY-TYPE TRANSFORMERS

Dry-type transformers are widely used because they are lighter in weight than comparable rated liquid-filled transformers. Installation is simpler because there is no need to take precautions against liquid leaks.

Dry-type transformers are constructed so that the core and coil are open to allow for cooling by the free movement of air. Fans may be installed to increase the cooling effect. In this case, the

High Voltage
Section

Transformer
Section

Secondary Distribution
Section

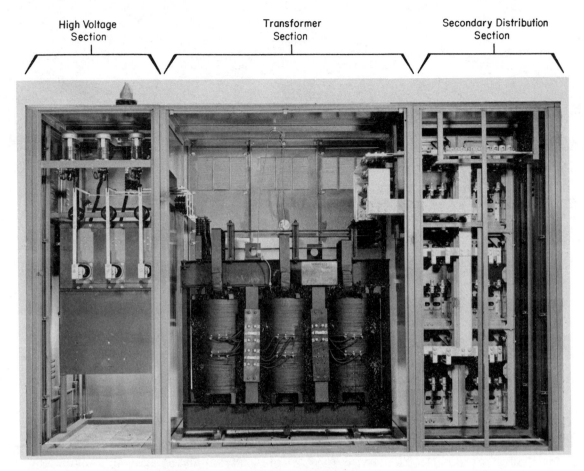

Fig. 2-1 A unit substation.

transformer can be used at a greater load level. A typical dry-type transformer installation is shown in figure 2-1. An installation of this type is known as a *unit substation* and consists of three main components: (1) the high-voltage switch, (2) the dry-type transformer, and (3) the secondary distribution section.

TRANSFORMER OVERCURRENT PROTECTION

Article 450 of the *NEC* covers transformer installations and groups transformers into two voltage levels:

　　1. Over 600 volts　　2. 600 volts or less

Figure 2-2, illustrates four of the more common situations found in commercial building transformer installations. As specified in *Article 450*, the overcurrent devices protect the transformer only. The conductors supplying or leaving the transformer may require additional overcurrent

protection according to *NEC Articles 240* and *310*. (See Units 18 and 19 of this text for information concerning fuses and circuit breakers.)

TRANSFORMER CONNECTIONS

A transformer is used in a commercial building primarily to reduce the transmission line high voltage to the value specified for the building. A number of connection methods can be used to accomplish the lowering of the voltage. The method used depends upon the requirements of the building. The following paragraphs describe several of the more commonly used secondary connection methods.

Single-phase System

Single-phase systems usually provide 120 and/or 240 volts with a two- or three-wire connection, figure 2-3. The center tap of the transformer secondary should be grounded in accordance with

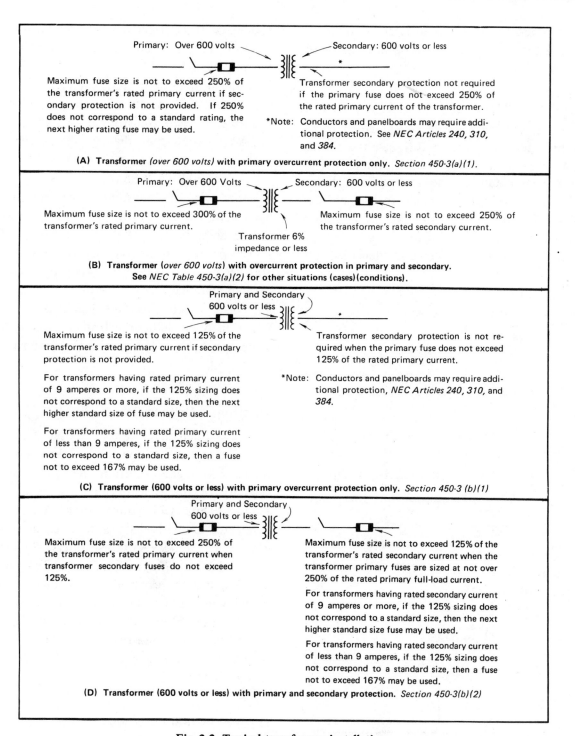

Primary: Over 600 volts   Secondary: 600 volts or less

Maximum fuse size is not to exceed 250% of the transformer's rated primary current if secondary protection is not provided. If 250% does not correspond to a standard rating, the next higher rating fuse may be used.

Transformer secondary protection not required if the primary fuse does not exceed 250% of the rated primary current of the transformer.

*Note: Conductors and panelboards may require additional protection. See NEC Articles 240, 310, and 384.

(A) Transformer (over 600 volts) with primary overcurrent protection only. Section 450-3(a)(1).

Primary: Over 600 Volts   Secondary: 600 volts or less

Maximum fuse size is not to exceed 300% of the transformer's rated primary current.

Maximum fuse size is not to exceed 250% of the transformer's rated secondary current.

Transformer 6% impedance or less

(B) Transformer (over 600 volts) with overcurrent protection in primary and secondary.
See NEC Table 450-3(a)(2) for other situations (cases)(conditions).

Primary and Secondary 600 volts or less

Maximum fuse size is not to exceed 125% of the transformer's rated primary current if secondary protection is not provided.

For transformers having rated primary current of 9 amperes or more, if the 125% sizing does not correspond to a standard size, then the next higher standard size of fuse may be used.

For transformers having rated primary current of less than 9 amperes, if the 125% sizing does not correspond to a standard size, then a fuse not to exceed 167% may be used.

Transformer secondary protection is not required when the primary fuse does not exceed 125% of the rated primary current.

*Note: Conductors and panelboards may require additional protection, NEC Articles 240, 310, and 384.

(C) Transformer (600 volts or less) with primary overcurrent protection only. Section 450-3 (b)(1)

Primary and Secondary 600 volts or less

Maximum fuse size is not to exceed 250% of the transformer's rated primary current when transformer secondary fuses do not exceed 125%.

Maximum fuse size is not to exceed 125% of the transformer's rated secondary current when the transformer primary fuses are sized at not over 250% of the rated primary full-load current.

For transformers having rated secondary current of 9 amperes or more, if the 125% sizing does not correspond to a standard size, then the next higher standard size fuse may be used.

For transformers having rated secondary current of less than 9 amperes, if the 125% sizing does not correspond to a standard size, then a fuse not to exceed 167% may be used.

(D) Transformer (600 volts or less) with primary and secondary protection. Section 450-3(b)(2)

Fig. 2-2 Typical transformer installations.

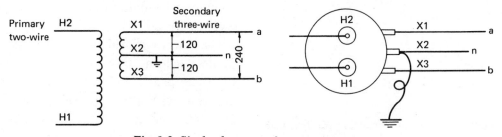

Fig. 2-3 Single-phase transformer connection.

Article 250, as will be discussed later. Grounding is a safety measure and should be installed with great care.

Three-phase Open Delta System

This connection has the advantage of being able to provide single-phase and three-phase power with the use of only two transformers. The grounded center tap of the transformer allows 120-volt connections to be made to two phases, figure 2-4. The third phase, however, is a *high leg* and cannot be used for lighting purposes. The high leg must be orange in color, or identified by

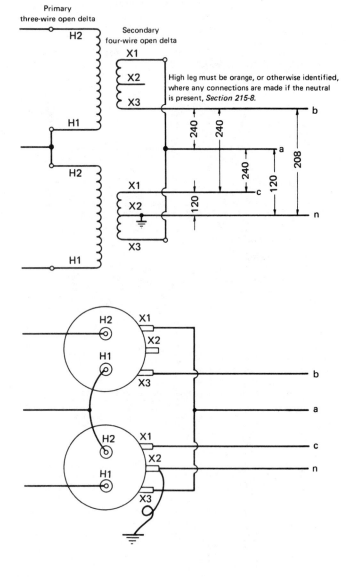

Fig. 2-4 Three-phase open delta connection for light and power.

tagging or other effective means where any connection is made if the neutral conductor is also present, *Section 215-8*.

In an open delta transformer bank, 86.6% of the capacity of the transformers is available. For example, if each transformer in figure 2-3 has a 100-kVA rating, then the capacity of the bank is:

$$100 + 100 = 200 \text{ kVA}$$
$$200 \text{ kVA} \times 86.6\% = 173 \text{ kVA}$$

Another way of determining the capacity of an open delta bank is to use 57.7% of the capacity of a full delta bank. Thus, the capacity of three 100-kVA transformers connected in full delta is 300 kVA. Two 100-kVA transformers connected in open delta have a capacity of

$$300 \times 57.7\% = 173 \text{ kVA}$$

When an open delta transformer bank is to serve three-phase power loads only, the center tap is not connected.

Three-phase Delta System

The open delta system of figure 2-4 can be closed by the addition of another transformer, figure 2-5. This addition increases the output capacity of the transformer bank by a factor of 1.732. Thus, while three 100-kVA transformers connected in delta have a total capacity of 300 kVA, two 100-kVA transformers connected in open delta have a capacity of only 173 kVA. Delta power is usually provided at 240 volts or 480 volts (other voltages may be used if necessary). Occasionally, the utility company will ground one phase of a three-phase delta system. This type of system is commonly called a *grounded B phase system*. (See Unit 18.)

Three-phase, Four-wire Wye System

The most commonly used system for modern commercial buildings is the three-phase, four-wire wye, figure 2-6. This system has the advantage of being able to provide three-phase power, and also permits lighting to be connected between any of the three phases and the neutral. Typical voltages available with this type of system are 120/208, 265/460, and 277/480. In each case, the transformer connections are the same.

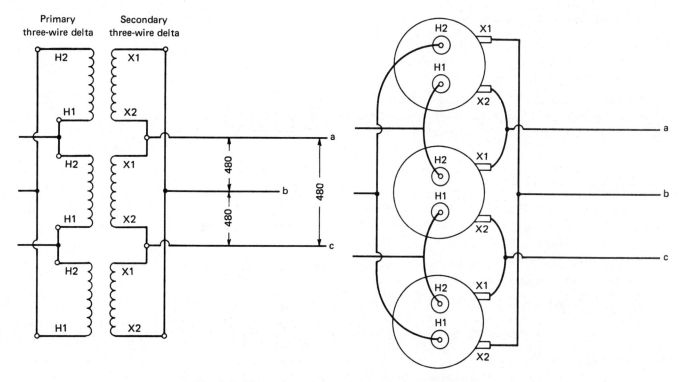

Fig. 2-5 Three-phase delta-delta transformer connection.

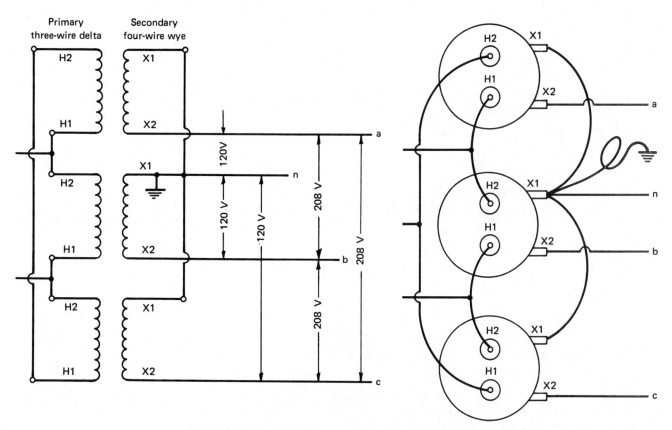

Fig. 2-6 Three-phase delta-wye transformer connection.

Notes to All Connection Diagrams

All connection diagrams are shown with additive polarity and standard angular displacement. All three-phase connections are shown with the primary connected in delta. For other connections, it is recommended that qualified engineering assistance be obtained.

THE SERVICE ENTRANCE

The regulations governing the method of bringing the electric power into a building are established by the local utility company. These regulations vary greatly between utility companies. Several of the more common methods of installing the service entrance are shown in figures 2-7 through 2-10.

Pad-mounted Transformers

Liquid-insulated transformers as well as dry-type transformers are used for this type of installation. These transformers may be fed by an underground or overhead service, figure 2-7. The secondary normally enters the building through a bus duct or large cable.

Unit Substation

For this installation, the primary runs directly to the unit substation where all of the necessary equipment is located, figure 2-8. The utility company requires the building owner to buy the equipment for a unit substation installation.

Pad-mounted Enclosure

Figure 2-9 illustrates an attractive arrangement in which the transformer and metering equipment are enclosed in a weatherproof cabinet. This type of installation (also known as a *transclosure*) is particularly adapted to smaller service-entrance requirements.

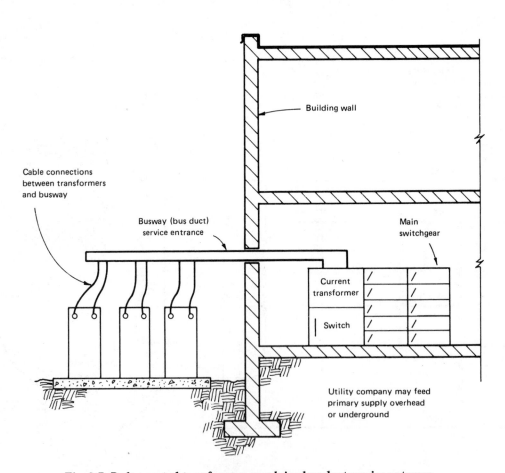

Fig. 2-7 Pad-mounted transformers supplying bus duct service entrance.

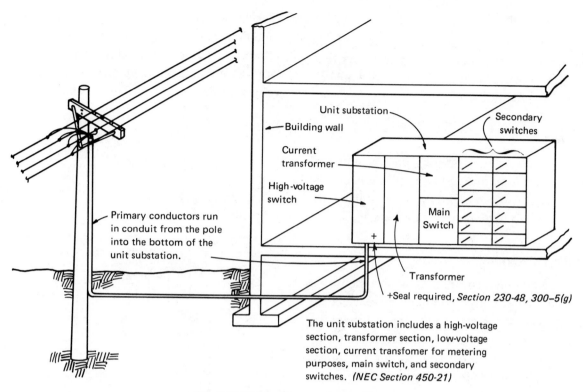

Unit substation

Building wall

Secondary switches

Current transformer

High-voltage switch

Main Switch

Primary conductors run in conduit from the pole into the bottom of the unit substation.

+

Transformer

+Seal required, *Section 230-48, 300–5(g)*

The unit substation includes a high-voltage section, transformer section, low-voltage section, current transformer for metering purposes, main switch, and secondary switches. *(NEC Section 450-21)*

Fig. 2-8 High-voltage service entrance.

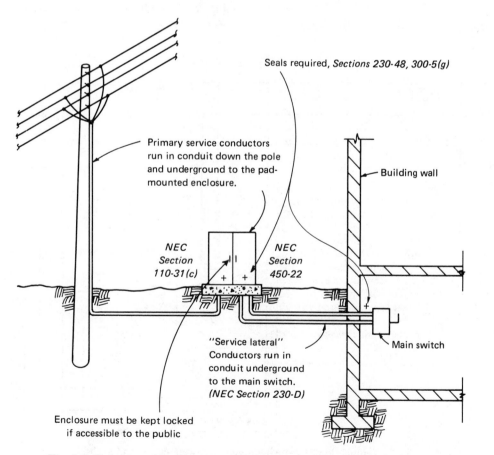

Seals required, *Sections 230-48, 300-5(g)*

Primary service conductors run in conduit down the pole and underground to the pad-mounted enclosure.

Building wall

NEC Section 110-31(c)

NEC Section 450-22

+ +

+

Main switch

"Service lateral" Conductors run in conduit underground to the main switch. *(NEC Section 230-D)*

Enclosure must be kept locked if accessible to the public

Fig. 2-9 Pad-mounted enclosure supplying underground service entrance.

Underground Vault

This type of service is used when available space is an important factor and an attractive site is desired. The metering may be at the utility pole or in the building, figure 2-10.

METERING

The electrician working on commercial installations seldom makes metering connections. However, the electrician should be familiar with the following two basic methods of metering.

High-voltage Metering

When a commercial building is occupied by a single tenant, the utility company may elect to meter the high-voltage side of the transformer. To accomplish this, a potential transformer and two current transformers are installed on the high-voltage lines and the leads are brought to the meter as shown in figure 2-11.

The left-hand meter socket in the illustration is connected to receive a standard socket-type watthour meter; the right-hand meter socket will receive a varhour meter (volt-ampere reactive meter). The two meters are provided with 15-minute demand attachments which register kilowatt (kW) and kilovolt-ampere reactive (kVAr) values respectively. These demand attachments

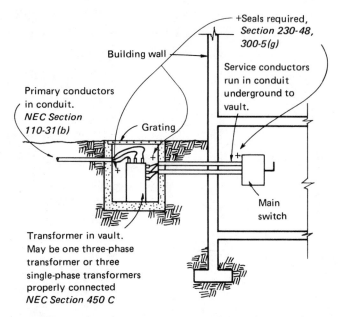

Fig. 2-10 Underground vault supplying underground service entrance.

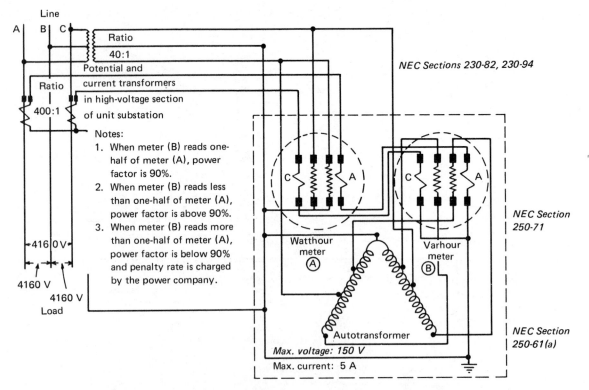

Fig. 2-11 Connections for high-voltage watthour meter and varhour meter. Demand attachments and timers are not shown.

will indicate the maximum usage of electrical energy for a 15-minute period during the interval between the readings made by the utility company. The rates charged by the utility company for electrical energy are based on the maximum demand and the power factor as determined from the two meters. A high demand or a low power factor will result in higher rates.

Low-voltage Metering

Low-voltage metering of loads greater than 200 amperes is accomplished in the same manner as high-voltage metering. In other words, potential and current transformers are used. For loads of 200 amperes or less, the feed wires from the primary supply are run directly to the meter socket, figure 2-12. For multiple occupancy buildings, such as

THREADED BOSS

LINE

LOAD

Fig. 2-12 Meter socket.

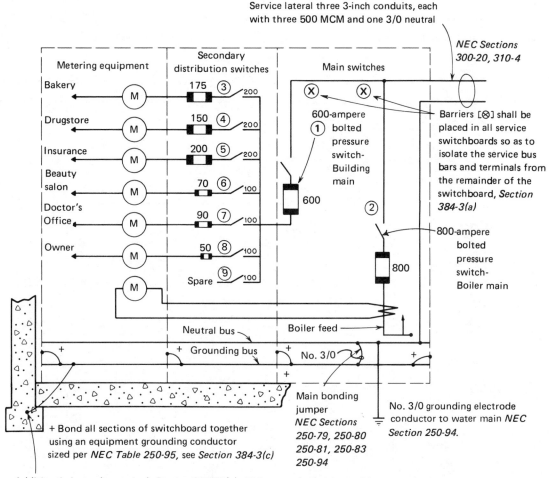

Service lateral three 3-inch conduits, each with three 500 MCM and one 3/0 neutral

NEC Sections 300-20, 310-4

Metering equipment

Secondary distribution switches

Main switches

Bakery — M — 175 ③ 200

Drugstore — M — 150 ④ 200

Insurance — M — 200 ⑤ 200

Beauty salon — M — 70 ⑥ 100

Doctor's Office — M — 90 ⑦ 100

Owner — M — 50 ⑧ 100

M — Spare ⑨ 100

600-ampere ① bolted pressure switch- Building main

600

② 800-ampere bolted pressure switch- Boiler main

800

Ⓧ Ⓧ

Barriers [⊗] shall be placed in all service switchboards so as to isolate the service bus bars and terminals from the remainder of the switchboard, *Section 384-3(a)*

Neutral bus

Boiler feed

Grounding bus

No. 3/0

Main bonding jumper *NEC Sections 250-79, 250-80 250-81, 250-83 250-94*

No. 3/0 grounding electrode conductor to water main *NEC Section 250-94.*

+ Bond all sections of switchboard together using an equipment grounding conductor sized per *NEC Table 250-95*, see *Section 384-3(c)*

Additional electrode required, *Section 250-81(a)*. This example shows an additional electrode buried in footing. Must be at least 20 feet (6.1 m) of bare solid copper not smaller than #4 AWG. In this case, #3/0 AWG is used, for convenience, because other 3/0 bare conductors are used for the bonding jumpers and grounding electrode conductor on this installation and thus were readily available at the project site.

Fig. 2-13 Commercial building service-entrance equipment, schematic drawing.

the commercial building investigated in this text, the meters are usually installed as a part of the service-entrance equipment.

SERVICE-ENTRANCE EQUIPMENT

When the transformer is installed at a location far from the building, the service-entrance equipment consists of the service-entrance conductors, the main switch or switches, the metering equipment, and the secondary distribution switches, figure 2-13. The commercial building shown in the plans is equipped in this manner and will be used as an example for the following paragraphs.

The Service

The service for the commercial building is similar to the service shown in figure 2-9. A pad-mounted, three-phase transformer is located outside the building and rigid conduit running underground serves as the service raceway.

Service Conductors

The total load in the commercial building is given as 1,135 amperes. Of this total, 91 amperes is used to operate the cooling system. This value of 91 amperes may be deducted from the total load due to the fact that the boiler and the cooling system do not operate at the same time and because the boiler requires 625 amperes. Therefore, for a load of 1,044 amperes, a good choice for the conductors is three No. 500 MCM type THHN or XHHW copper conductors run in three-inch parallel conduits. (Refer to *Table 310-16* of the *NEC.*) The calculated neutral load is 403 amperes; thus, the size of the neutral can be reduced. To minimize induced currents, the same size conductors must be installed in each raceway for each phase (see *Section 300-20* of the *NEC*). A No. 3/0 type THHN or XHHW copper conductor is used for the neutral. There will be one 3/0 conductor in each of the three 3-inch conduits. The service-entrance equipment is shown in figure 2-14.

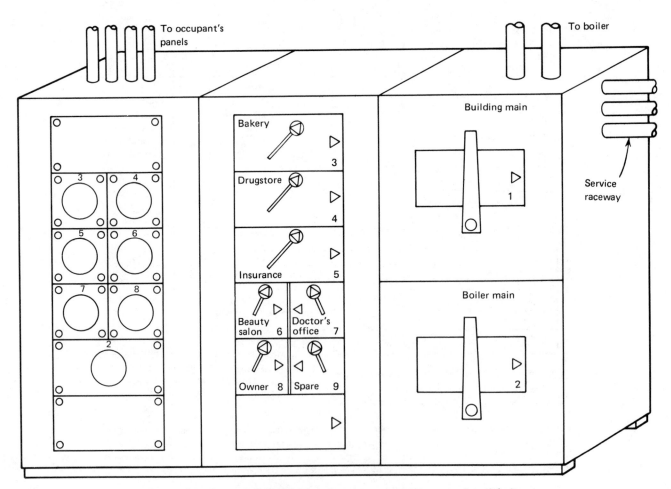

Fig. 2-14 Commercial building service-entrance equipment, pictorial view.

The service lateral will be tapped in the main gear enclosure to feed one 800-ampere and one 600-ampere fusible main switch. (These switches are shown in figures 2-13 and 2-15.) The 800-ampere main switch serves the boiler. The 600-ampere switch serves the secondary distribution switches consisting of three 200-ampere switches and four 100-ampere switch units. Fuse reducers installed in the switch will accommodate the 50-ampere fuses for the feeder to the owner's panelboard.

A summary of the service and feeder calculations is given in Table 2-1. The details of the calculations used to arrive at the values given are covered later in this text in the appropriate units.

Throughout the design of the electrical system used in the commercial building in the plans, a simple method is used to calculate the required sizes of the branch circuit, feeder, and service-entrance conductors. The method used includes the following considerations:

- the general lighting load calculations based on the number of watts per square foot, the motor load, and any other loads.

- in calculating single-phase loads, the product of 120×1.73 is rounded to 208 and, when calculating three-phase loads, the product of 208×1.73 is rounded to 360.

- an allowance of 25% for future growth.

- an allowance of 25% for the derating of the conductor ampacity as required by *Notes 8* and *10* to *Table 310-16* of the *NEC.*

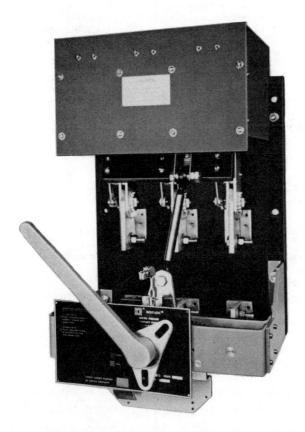

(A) Bolted pressure contact switch for use with Class J, Class L, or Class T high-capacity fuses. The service for the commercial building in the plans has two of these switches as the main disconnecting means. (Figures 2-13, 2-14)

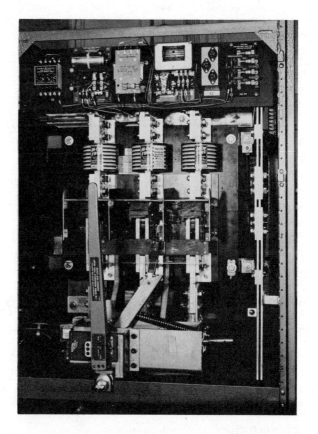

(B) These switches can be purchased with ground-fault protection, shunt tripping, auxiliary contacts, phase failure relays, and other optional attachments. Switch illustrated has ground-fault protection, phase failure relay, shunt tripping, and antisingle phasing blown fuse indicator.

Fig. 2-15

TABLE 2-1 ELECTRICAL DESIGN SUMMARY

Occupancy	a. Loading (watts)	φ	b. Growth (watts)	c. Derating (watts)	d. Total (watts)	Ampere	Conductor phase	Conductor neutral	Conduit Size	① Am-pacity	② Over-current Protec-tion	Switch Size
Drugstore	27,580	3 φ 4 W	6,895	8,619	43,094	120	1 ③	4	1 1/4"	150	150	200
Bakery	33,900	3 φ 4 W	8,475	10,593	52,968	147	1/0	6	1 1/2"	170	175	200
Insurance Office	46,512	3 φ 4 W	11,628	14,535	72,675	202	3/0	2/0	2"	225	200 ④	200
Beauty Salon	15,511	3 φ 4 W	3,878	4,847	24,236	67	6	8	1"	75	70	100
Doctor's Office	11,738	1 φ 3 W	2,935	⑦	14,673	71	4	4	1"	95	90	100
Boiler	225,000	3 φ 3 W	—	56,250	281,250	781	2-500	—	2-3"	860	800 ⑥	800
Owner	11,319	3 φ 4 W	2,830	3,537	17,686	49	8	8	1"	55	50	100 ⑤
Building Summary	371,560	3 φ 4 W	36,641	⑧	408,201 e.	1,135 f.	3-500	3-3/0	3-3"	1,290	800 + 600	800+ 600

a. Loading (watts) = the calculated load as per Code requirements.
b. Growth (watts) = Loading watts × 0.25
c. Derating (watts) = (Loading watts + Growth watts) × 0.25
d. Total (watts) = Loading (watts) × 1.25 × 1.25
               or
  Loading (watts) + Growth (watts) + Derating (watts)
e. 408,201 = 371,560 + 36,641
f. Amperes = $\dfrac{\text{Watts}}{1.73 \times 208} = \dfrac{408,201}{1.73 \times 208} = 1{,}135$ amperes

---

1. Conductor types THHN or XHHW. Ampacities taken directly from *NEC Table 310-16*.
2. Use the next larger size, see *NEC Sections 240-3* and *240-6*.
3. A No. 2 AWG conductor would carry the load but a No. 1 AWG conductor was chosen as it provides more capacity without changing the conduit or switch size.
4. Although less than the calculated ampacity requirements, the two-ampere additional capacity would require a 400-ampere switch. Common sense should prevail when this type of selection is made.
5. A 100-ampere switch is specified because the units are paired and a 100-ampere spare is preferred.
6. See *Section 240-3, Exception 1;* the selection of a protective device with the next higher rating is not permitted over 800 amperes.
7. Derating is not required since only three conductors are installed in the raceway.
8. Derating is not required, neutral need not be counted, as only small portion of total load is electric-discharge lighting.

Thus,

$$\boxed{\begin{array}{c}\text{recommended size}\\ \text{for branch circuit,}\\ \text{feeder service}\end{array}} = \boxed{\begin{array}{c}\text{calculated}\\ \text{loading}\end{array}} \times \boxed{1.25} \times \boxed{1.25}$$

The metering for the service entrance of this commercial installation is located in the third section of the main service-entrance equipment. The meters for six of the occupancies in the building are in-stalled directly in the line. The seventh meter and current transformers are installed to measure the load to the boiler. The service equipment can be purchased in a single enclosure, such as the enclosure illustrated for the commercial building in this text. The contractor and electrician should consult the *National Electrical Code* and local regulations, and obtain the utility company's approval of location and installation requirements before installing any service.

GROUNDING

Why is grounding necessary?

Electrical systems and their associated conductors are grounded to keep voltage spikes to a minimum should lightning hit or whenever other line surges occur. Grounding stabilizes normal voltage to ground.

Electrical metallic conduits and equipment are grounded to keep the voltage to ground on equipment to a minimum, thus reducing the hazard of shock.

It is very important that the ground connections and the grounded system be properly installed. To achieve the best possible ground system, the electrician must use the recommended procedures when installing the proper equipment. Figure 2-16 illustrates some of the terminology used in *NEC Article 250, Grounding.*

Properly grounded electrical systems and enclosures provide a good path for ground currents if a ground fault occurs or if the insulation fails. The reason for this is that the lower the impedance of the grounding path (ac resistance to the current), the greater is the ground-fault current. This increased ground-fault current causes the overcurrent device protecting the circuit to respond faster than it otherwise would for low-value currents. For example,

$$I = \frac{E}{Z} = \frac{277}{0.1} = 2{,}770 \text{ amperes}$$

but, with a lower impedance,

$$I = \frac{E}{Z} = \frac{277}{0.01} = 27{,}700 \text{ amperes}$$

The amount of ground-fault damage to electrical equipment is related to (1) the response time of the overcurrent device and (2) the amount of current. One common term used to relate the time and current to the ground-fault damage is *ampere squared seconds* ($I^2 t$):

Amperes $\times$ Amperes $\times$ Time in seconds = $I^2 t$

It can be seen in the expression for $I^2 t$ that when the current (I) and time (t) in seconds are kept to a minimum, a low value of $I^2 t$ results. Lower values of $I^2 t$ mean that less ground-fault damage will occur. Units 18 and 19 provide detailed coverage of overcurrent protective devices, fuses, and circuit breakers.

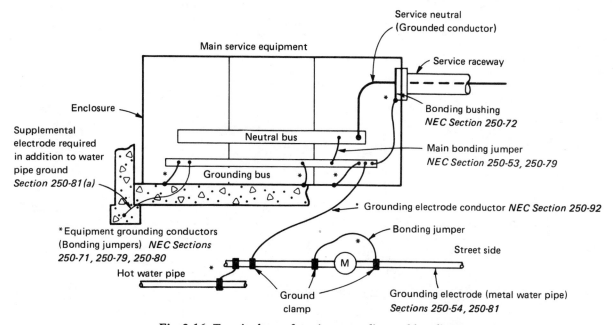

Fig. 2-16 Terminology of service grounding and bonding.

System Grounding

System grounding calls for grounding of the entire system rather than grounding of a single item. By system is meant, for example, the service neutral conductor, hot and cold water pipes, gas pipes, service-entrance equipment and jumpers around meters. If any of these system parts become disconnected or open in any way, the integrity of the system is still maintained through the other paths. In other words, everything is tied together. The concept of system grounding is shown in figure 2-17.

Figure 2-17 and the following steps illustrate what can happen if an entire system is *not* grounded.

1. A live wire contacts the gas pipe. The bonding jumper (A) is not installed originally.
2. The gas pipe now has 120 volts through it.
3. The insulating joint in the gas pipe results in a poor path to ground; assume the resistance is 8 ohms.
4. The 20-ampere fuse does not blow.

$$I = \frac{E}{R} = \frac{120}{8} = 15 \text{ amperes}$$

5. If a person touches the hot gas pipe and the water pipe at the same time, current flows through the person's body. If the body resistance is 12,000 ohms, then the current is:

$$I = \frac{E}{R} = \frac{120}{12,000} = 0.01 \text{ ampere}$$

This value of current passing through a human body can cause death.

6. The fuse is now "seeing" 15 + 0.01 = 15.01 amperes; however, it still does not blow.
7. If the system grounding concept had been used, bonding jumper (A) would have kept the voltage difference between the water pipe and the gas pipe at zero. Thus, the fuse would blow. If 10 feet (3.05 m) of No. 4 AWG copper wire were used as the jumper, then the resistance of the jumper is 0.00259 ohm. The current is:

$$I = \frac{E}{R} = \frac{120}{0.00259} = 46,332 \text{ amperes}$$

(In an actual system, the impedance of all of the parts of the circuit would be much higher. Thus, a much lower current would result. The value of current, however, would be enough to cause the fuse to blow.)

Advantages of System Grounding

The advantages of system grounding are:

- the potential voltage differentials between the different parts of the system are kept at a minimum, thereby reducing shock hazard.

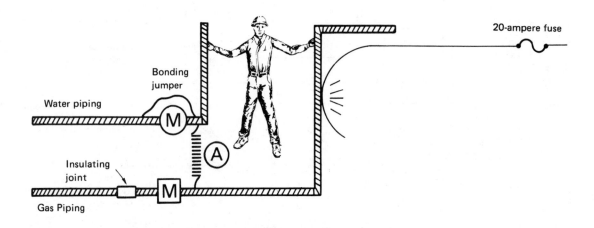

Fig. 2-17 System grounding.

• impedance of ground path is kept at a minimum which results in higher current flow in the event of a ground fault. The lower the impedance, the greater the current flow, and the faster the overcurrent device opens.

In figures 2-13 and 2-16, the main service equipment, the service raceways, the neutral bus, the grounding bus, and the hot and cold water pipes have been bonded together to form a *Grounding Electrode System*, as required in *Section 250-81* of the Code.

For discussion purposes, at least 20 feet (6.1 m) of No. 3/0 bare copper conductor is installed in the footing to serve as the additional supplemental electrode that is required by *Section 250-81* of the Code. The minimum size permitted in No. 4 copper. This conductor, buried in the concrete footing, is supplemental to the water pipe ground. *Section 250-81(a)* permits the supplemental electrode to be bonded to the grounding electrode conductor, the grounded service-entrance conductor, the grounded service raceway or the interior metal water piping at any convenient point.

Had No. 3/0 bare copper conductor not been selected for installation in the concrete footing, the supplemental grounding electrode could have been:

• the metal frame of the building.

• at least 20 feet (6.1 m) of steel reinforcing bars (re-bars) 1/2-inch (12.7 mm) minimum diameter, encased in concrete at least 2 inches (50.8 mm) thick, in direct contact with the earth, such as near the bottom of the foundation or footing.

• at least 20 feet (6.1 m) of bare copper wire, encircling the building, minimum size No. 2 AWG, buried directly in the earth at least 2 1/2 feet (762 mm) deep.

• the gas pipe main.

• ground rods.

• ground plates.

All of these factors are discussed in *NEC Article 250, Parts G, H, J,* and *K.*

Many Code interpretations are probable as a result of the grounding electrode systems concept. Therefore, the local code authority should be consulted. For instance, some electrical inspectors may not require the bonding jumper between the hot and cold water pipes, as shown in figure 2-16. They may determine that an adequate bond is provided through the water heater itself of either type, electric or gas. Other electrical inspectors may require that the hot and cold water pipes be bonded together because some water heaters contain insulating fittings that are intended to reduce corrosion inside the tank caused by electrolysis. According to their reasoning, even though the water heater originally installed contains no fittings of this type, such fittings may be included in a future replacement heater, thereby necessitating a bond between the cold and hot water pipes. The local electrical inspector should be consulted for the proper requirements. However, bonding the pipes together, as shown in figure 2-16, is the recommended procedure.

Code Grounding Rules (*Article 250*)

When grounding service-entrance equipment, the following Code rules must be observed:

• The system must be grounded when maximum voltage to ground does not exceed 150 volts, *Section 250-5(b)(1).*

• All grounding schemes shall be installed so that no objectionable currents will flow over the grounding conductors and other grounding paths, *Section 250-21(a).*

• The grounding electrode conductor must be connected to the supply side of the service disconnecting means, *Section 250-23(a).*

• The identified neutral conductor is the conductor that must be grounded, *Section 250-25.*

• Tie (bond) everything together, *Sections 250-80(a)* and *(b),* and *250-81.*

• The grounding electrode conductor used to connect the grounded neutral conductor to the grounding electrode must not be spliced, *Section 250-53.*

• The grounding conductor is to be sized as per *Table 250-94.*

Table 250-94
Grounding Electrode Conductor for AC Systems

Size of Largest Service-Entrance Conductor or Equivalent Area for Parallel Conductors		Size of Grounding Electrode Conductor	
Copper	Aluminum or Copper-Clad Aluminum	Copper	*Aluminum or Copper-Clad Aluminum
2 or smaller	0 or smaller	8	6
1 or 0	2/0 or 3/0	6	4
2/0 or 3/0	4/0 or 250 MCM	4	2
Over 3/0 thru 350 MCM	Over 250 MCM thru 500 MCM	2	0
Over 350 MCM thru 600 MCM	Over 500 MCM thru 900 MCM	0	3/0
Over 600 MCM thru 1100 MCM	Over 900 MCM thru 1750 MCM	2/0	4/0
Over 1100 MCM	Over 1750 MCM	3/0	250 MCM

Where there are no service-entrance conductors, the grounding electrode conductor size shall be determined by the equivalent size of the largest service-entrance conductor required for the load to be served.

* See installation restrictions in Section 250-92(a).

See Section 250-23(b).

Reprinted with permission from NFPA 70-1984, *National Electrical Code®*, Copyright © 1983, National Fire Protection Association, Quincy, Massachusetts 02269. This reprinted material is not the complete and official position of the NFPA on the referenced subject, which is represented only by the standard in its entirety.

- The hot and cold water metal piping system shall be bonded to the service equipment enclosure, to the grounded conductor at the service, and to the grounding electrode conductor, *Section 250-80*.

- The grounding electrode conductor must be connected to the metal underground water pipe when 10 feet (3.05 m) long or longer, including the well casing, *Section 250-81*.

- In addition to grounding the service equipment to the underground water pipe, one or more additional electrodes are required, such as a bare conductor in the footing, *Section 250-81(c)*; a grounding ring, *Section 250-81(d)*; a metal underground gas piping system, *Section 250-83(a)*; rod or pipe electrodes, *Section 250-83(c)*; or plate electrodes, *Section 250-83(d)*. All of these items must be bonded together if available on the premises, *Section 250-81*.

- The grounding conductor shall be copper, aluminum or copper-clad aluminum, *Section 250-91*.

- The grounding conductor may be solid or stranded, uninsulated, covered or bare, and must not be spliced, *Section 250-91*.

- Bonding is required around all insulating joints or sections of the metal piping system that might be disconnected, *Section 250-112*.

- The connection to the grounding electrode must be accessible, *Section 250-112*.

- The grounding conductor must be tightly connected by using proper lugs, connectors, clamps or other approved means, *Section 250-115*. One type of grounding clamp is shown in figure 2-18.

Sizing the Grounding Electrode Conductors

The grounding electrode conductor connects the grounding electrode to the equipment grounding conductor and to the systems grounding conductor. In the commercial building, this means that the grounding electrode conductor connects the main water pipe *(grounding electrode)* to the grounding bus *(equipment grounding conductor)* and to the neutral bus *(systems grounding conductor)*. See figure 2-16.

Table 250-94 is referred to when selecting grounding electrode conductors for services where there is overcurrent protection ahead of the service-entrance conductors other than the utility companies' primary or secondary overcurrent protection.

In *Table 250-94* of the *National Electrical Code,* the service conductor sizes are given in the wire size and not the ampacity values. To size a grounding electrode conductor for a service with three parallel No. 500 MCM service conductors, it is necessary to total the wire size and select a grounding electrode conductor for an equivalent No. 1500 MCM service conductor. In this case, the grounding electrode conductor is No. 3/0 copper (see figure 2-13). Figure 2-19 shows another example.

Refer to figures 2-20 through 2-22 for the proper procedures to be used in bonding electrical equipment.

Table 250-95 of the Code is referred to when selecting equipment grounding conductors when there is overcurrent protection ahead of the conductor supplying the equipment.

NEC Table 250-95 is based on the current setting or rating, in amperes, of the overcurrent device installed ahead of the equipment (other than service-entrance equipment) being supplied.

The electric boiler feeder in the commercial building consists of two No. 500 MCM conductors per phase protected by 800-ampere fuses in the main

Table 250-95. Minimum Size Equipment Grounding Conductors for Grounding Raceway and Equipment

Rating or Setting of Automatic Overcurrent Device in Circuit Ahead of Equipment, Conduit, etc., Not Exceeding (Amperes)	Size	
	Copper Wire No.	Aluminum or Copper-Clad Aluminum Wire No.*
15	14	12
20	12	10
30	10	8
40	10	8
60	10	8
100	8	6
200	6	4
300	4	2
400	3	1
500	2	1/0
600	1	2/0
800	0	3/0
1000	2/0	4/0
1200	3/0	250 MCM
1600	4/0	350 "
2000	250 MCM	400 "
2500	350 "	600 "
3000	400 "	600 "
4000	500 "	800 "
5000	700 "	1200 "
6000	800 "	1200 "

*See installation restrictions in Section 250-92(a)
Reprinted with permission from NFPA 70-1984, *National Electrical Code®,* Copyright © 1983, National Fire Protection Association, Quincy, Massachusetts 02269. This reprinted material is not the complete and official position of the NFPA on the referenced subject, which is represented only by the standard in its entirety.

Fig. 2-18 Grounding clamp.

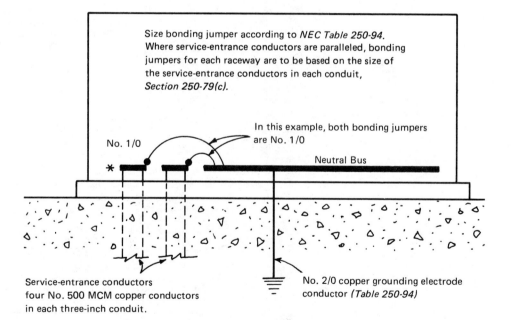

Fig. 2-19 Typical service entrance (grounding and bonding).

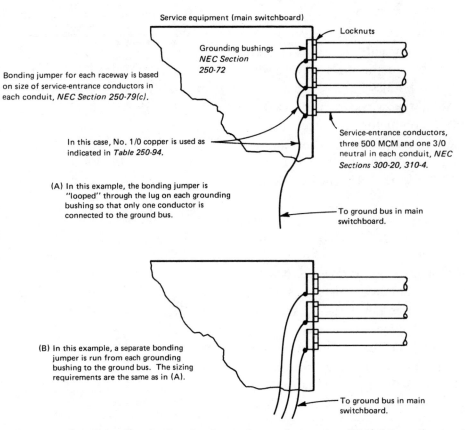

Service equipment (main switchboard)

Locknuts

Grounding bushings
NEC Section
250-72

Bonding jumper for each raceway is based
on size of service-entrance conductors in
each conduit, NEC Section 250-79(c).

Service-entrance conductors,
three 500 MCM and one 3/0
neutral in each conduit, NEC
Sections 300-20, 310-4.

In this case, No. 1/0 copper is used as
indicated in Table 250-94.

(A) In this example, the bonding jumper is
"looped" through the lug on each grounding
bushing so that only one conductor is
connected to the ground bus.

To ground bus in main
switchboard.

(B) In this example, a separate bonding
jumper is run from each grounding
bushing to the ground bus. The sizing
requirements are the same as in (A).

To ground bus in main
switchboard.

Fig. 2-20 Sizing of main bonding jumper for the service on the commercial building. *Section 250-79.*

(A) On service, a bonding jumper shall be used around concentric and eccentric knockouts, *NEC Section 250-72(d).*

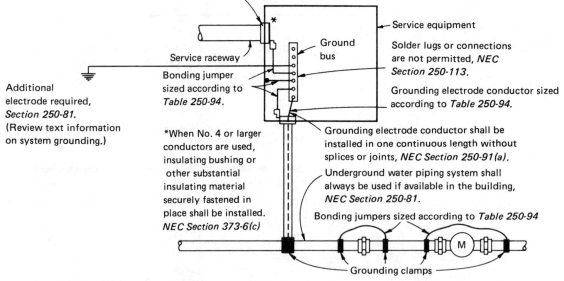

Service equipment

Ground
bus

Service raceway

Solder lugs or connections
are not permitted, *NEC
Section 250-113.*

Bonding jumper
sized according to
Table 250-94.

Grounding electrode conductor sized
according to *Table 250-94.*

Additional
electrode required,
Section 250-81.
(Review text information
on system grounding.)

Grounding electrode conductor shall be
installed in one continuous length without
splices or joints, *NEC Section 250-91(a).*

*When No. 4 or larger
conductors are used,
insulating bushing or
other substantial
insulating material
securely fastened in
place shall be installed.
NEC Section 373-6(c)*

Underground water piping system shall
always be used if available in the building,
NEC Section 250-81.

Bonding jumpers sized according to *Table 250-94*

M

Grounding clamps

Bonding is provided to insure that the grounding will conduct safely any fault current likely to be imposed on
the system. *NEC Section 250-70.*

(B)

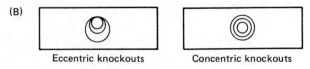

Eccentric knockouts

Concentric knockouts

Fig. 2-21 Bonding and grounding of service equipment.

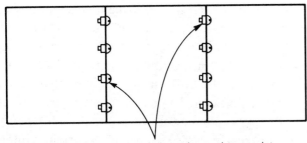

Nonconductive paint, enamel, or similar coating must be removed at contact points when sections of electrical equipment are bolted together to insure that the sections are effectively bonded. Tightly driven metal bushings and locknuts generally will bite through paints and enamels, thus making the removal of the paint unnecessary. *NEC Section 250-75.*

Fig. 2-22 Insuring bonding when electrical equipment is bolted together.

switchboard. The use of flexible metal conduit at the boiler means that a bonding wire must be run through each conduit as shown in the figure. A detailed discussion on flexible connections is given in Unit 8. An illustration of a typical application for the electric boiler feeder in the commercial building is shown in figure 2-23.

GROUND-FAULT PROTECTION

The *NEC* requires the use of ground-fault protection (GFP) devices on services which meet the conditions outlined in *NEC Section 230-95.* Thus, ground-fault protection devices are installed:

- on solidly grounded wye services above 150 volts to ground, but not over 600 volts

between phases (for example, on 277/480-volt systems).

- on service disconnects rated at 1,000 amperes or more.

- to operate at 1,200 amperes or less.

- so that the maximum time of opening the service switch or circuit breaker does not exceed one (1) second for ground-fault currents of 3,000 amperes or more.

- to limit damage to equipment and conductors on the *load side* of the service disconnecting means. GFP will *not* protect against damage caused by faults occurring on the *line side* of the service disconnect.

These ground-fault protection requirements do not apply on services for a continuous process where a nonorderly shutdown will introduce additional or increased hazards, *Sections 230-95 and 240-12.*

When a fuse-switch combination serves as the service disconnect, the fuses must have adequate interrupting capacity to interrupt the available fault current (*Section 110-9*), and must be capable of opening any fault currents that exceed the interrupting rating of the switch during any time when the ground-fault protective system will not cause the switch to open, *Section 230-95(b).*

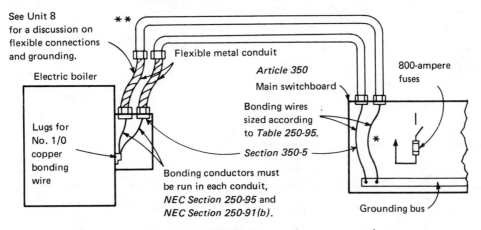

*Boiler feeder consists of two No. 500 MCM copper conductors connected in parallel (two phase).

** Each conduit contains three No. 500 MCM "hot" conductors and one No. 1/0 bonding (equipment ground) conductor solidly connected to the grounding bus in the switchboard and to the lugs in the terminal box on the boiler. The No. 1/0 bonding conductor may be insulated or bare, *NEC Section 250-91(b).*

Fig. 2-23 Sizing equipment grounding conductors for the electric boiler feeder.

Ground-fault protection is not required on:

- delta-connected three-phase systems.

- ungrounded wye-connected three-phase systems.

- single-phase systems.

- 120/240-volt single-phase systems.

- 120/208-volt three-phase systems.

- systems over 600 volts; for example, 2,400/4,160 volts.

- service disconnecting means rated at less than 1,000 amperes.

- systems where the service is subdivided; for example, a 1,600-ampere service may be divided between two 800-ampere switches.

The time of operation of the device as well as the ampere setting of the GFP device must be considered carefully to insure that the continuity of the electrical service is maintained. The time of operation of the device includes: (1) the sensing of a ground fault by the GFP monitor, (2) the monitor signaling the disconnect switch to open, and (3) the actual opening of the contacts of the disconnect device (either a switch or a circuit breaker). The total time of operation may result in a time lapse of several cycles or more (Units 18 and 19).

GFP circuit devices were developed to overcome a major problem in circuit protection: the low-value phase-to-ground arcing fault, figure 2-24. The amount of current that flows in an arcing phase-to-ground fault can be low when compared to the rating or setting of the overcurrent device. For example, an arcing fault can generate a current

flow of 600 amperes. Yet, a main breaker rated at 1,600 amperes will allow this current to flow without tripping, since the 600-ampere current appears to be just another *load* current. The operation of the GFP device assumes that under normal conditions the total instantaneous current in all of the conductors of a circuit will exactly balance, figure 2-25. Thus, if a current coil is installed so that all of the circuit conductors run through it, the normal current measured by the coil will be zero. If a ground fault occurs, some current will return through the grounding system and an unbalance will result in the conductors. This unbalance is then detected by the GFP device, figure 2-26.

The purpose of ground-fault protection devices is to sense and protect equipment against *low-level ground faults*. GFP monitors do not sense phase-to-phase faults, three-phase faults, or phase-to-neutral faults. These monitors are designed to sense phase-to-ground faults only.

Large (high magnitude) ground-fault currents can cause destructive damage even though a GFP is

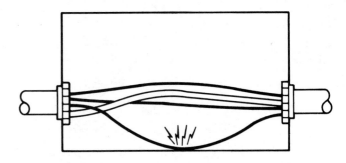

Arcing ground fault can occur where a phase wire and the conduit grounding system contact each other.

Fig. 2-24 Ground faults.

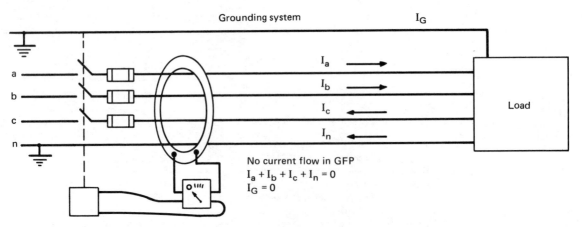

Fig. 2-25 Normal condition.

installed. The amount of arcing damage depends upon (1) how much current flows and (2) the length of time that the current exists. For example, if a GFP device is set for a ground fault of 500 amperes and the time setting is six cycles, then the device will need six cycles to signal the switch or circuit breaker to open the circuit whether the ground fault is slightly more than 500 amperes or as large as 20,000 amperes. The six cycles needed to signal the circuit breaker plus the operation time of the switch or breaker may be long enough to result in damage to the switchgear.

The damaging effects of high magnitude ground faults, phase-to-phase faults, three-phase faults, and phase-to-neutral faults can be reduced substantially by the use of current-limiting overcurrent devices. These devices reduce both the peak let-through current and the time of opening once the current is sensed. For example, a ground fault of 20,000 amperes will open a current-limiting fuse in less than one-half cycle. In addition, the peak let-through current is reduced to a value much less than 20,000 amperes (see Units 18 and 19).

Ground-fault protection for equipment, *NEC Section 210-8*, is not to be confused with personnel ground-fault protection (GFCI or GFI). (GFCI protection is covered in detail in *Electrical Wiring — Residential.*)

The GFP is connected to the normal fused switch or circuit breaker which serves as the circuit protective device. The GFP is adjusted so that it will signal the protective device to open under abnormal ground-fault conditions. The maximum setting of the GFP is 1,200 amperes. In the commercial building in the plans, the service voltage is 120/208 volts. The voltage to ground on this system is 120 volts. This value is not large enough to sustain an electrical arc. Therefore, it is not required (according to the *NEC*) that ground-fault protection for the service disconnecting means be installed for the commercial building. The electrician can follow a number of procedures to minimize the possibility of an arcing fault. Examples of these procedures follows.

- Insure that conductor insulation is not damaged when the conductors are pulled into raceway.
- Insure that the electrical installation is properly grounded and bonded.
- Locknuts and bushings must be tight.
- All electrical connections must be tight.
- Tightly connect bonding jumpers around concentric and/or eccentric knockouts.
- Insure that conduit couplings and other fittings are installed properly.
- Check insulators for minute cracks.
- Install insulating bushings on all raceways.
- Insulate all bare bus bars in switchboards when possible.

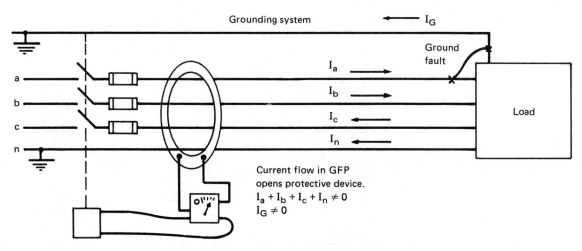

Fig. 2-26 Abnormal condition.

- Conductors must not rest on bolts, or other sharp metal edges.

- Electrical equipment must not be allowed to become damp or wet either during or after construction.

- All overcurrent devices must have an adequate interrupting capacity.

- Do not work on *hot* panels.

- Be careful when using *fish tape* since the loose end can become tangled with the electrical panelboard.

- Be careful when working with *live* parts; do not drop tools on top of such parts.

- Avoid large services; for example, it is usually preferable to install two 800-ampere service disconnecting means rather than to install one 1,600-ampere service disconnecting means.

REVIEW

Note: Refer to the *National Electrical Code* or the plans as necessary.

1. A 300-kVA, dry-type transformer bank has a three-phase, 480-volt delta primary and a three-phase, 208-volt wye secondary. What is the proper size of fuse, in amperes, that must be installed in the secondary? _____

2. Draw the secondary connections for a three-phase, four-wire wye-connected transformer bank. Label the locations where the voltage across a transformer is 120 volts.

3. A panelboard is added to an existing panelboard installation. Two knockouts, one near the top of the panelboard and the other near the bottom, are cut in the adjoining sides of the boxes. Indicate on the drawing the proper way of extending the phase and neutral conductors to the new panelboard. Line side lugs are suitable for two conductors.

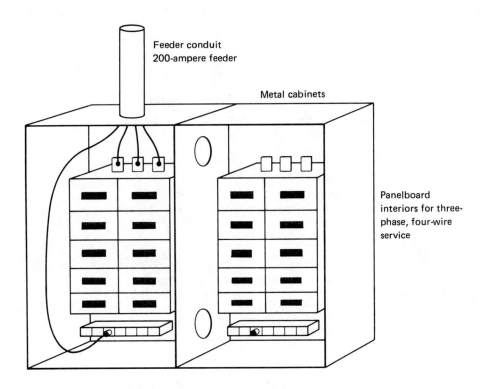

Feeder conduit
200-ampere feeder

Metal cabinets

Panelboard
interiors for three-
phase, four-wire
service

4. List any five of the procedures that can be followed to minimize the possibility and/or severity of an arcing fault.

_____

_____

_____

_____

_____

5. A five-horsepower, three-phase, 230-volt motor is installed. The motor requires the installation of an equipment grounding conductor in the same conduit as the motor branch circuit. The full-load rating of the motor is 15.2 amperes. The circuit consists of No. 12 TW wire with an ampacity of 20 amperes. However, the circuit is provided with 30-ampere short-circuit protection. What size copper conductor is used for the equipment grounding conductor? _____

6. The service of the commercial building consists of three _____-inch conduits, each containing three No. _____ MCM phase conductors plus one No. _____ neutral conductor.

7. What is the purpose of grounding systems and enclosures?

_____

_____

_____

8. What is the kVA-capacity of each of the following. (Underline the correct answer.)
   a. Three 75-kVA transformers connected to closed delta.
      (1) 225 kVA   (2) 150 kVA   (3) 130 kVA
   b. Two 75-kVA transformers connected in open delta.
      (1) 225 kVA   (2) 150 kVA   (3) 130 kVA
   c. Three 150-kVA transformers wye connected.
      (1) 300 kVA   (2) 450 kVA   (3) 225 kVA

9. Of the following, which is the proper size copper grounding electrode conductor for a 100-ampere service that consists of No. 3/0 phase conductors?
   a. No. 4 AWG
   b. No. 2 AWG
   c. No. 6 AWG

10. A service is supplied by three No. 350 MCM conductors per phase. What is the minimum size copper grounding electrode conductor to be used?
   a. No. 1/0 AWG
   b. No. 2/0 AWG
   c. No. 3/0 AWG

11. Ground-fault sensing equipment is required on certain types of services. For the following systems, indicate where this equipment is used by inserting in the blanks either R (Required) or N (Not Required).
   a. 120/208-volt system   _____
   b. 277/480-volt system   _____
   c. delta systems   _____
   d. single-phase systems   _____
   e. 1,200-ampere service, three-phase/four-wire wye, 277/480 volts   _____
   f. 800-ampere service   _____

12. The Code requires that if a metal water piping system is available on the premises, the electrical system must be grounded to this water piping system. In addition to the water pipe ground, the Code requires at least _____ additional ground. Name the additional types of grounds permitted.

_____

_____

_____

_____

_____

_____

_____

_____

_____

_____

UNIT 3

Reading Electrical Drawings (Prints)—Drugstore

OBJECTIVES

After completing the study of this unit, the student will be able to

- identify symbols associated with the reading of electrical drawings.
- determine the requirements of the electrical contract.
- calculate plan dimensions.

PRINTS

Electrical drawings are the maps which the electrician must read and understand to be able to install a complete electrical system and coordinate related activities with those of workers in the other crafts involved in the construction project. Using a standard system of symbols, these electrical prints guide the construction of a building; therefore, the electrician must acquire the ability to read the various symbols appearing on the prints as they apply to the electrical work.

The several units in this text which deal with electrical drawings and plans will apply the information presented in units prior to the specific print unit. The questions at the end of each print unit will require the student to use the specifications and the prints to obtain the necessary information.

APPLICATION QUESTIONS

Answer questions 1–7 by identifying the symbols and indicating the number of installations shown in the drugstore wiring system for each item. Refer to the *National Electrical Code* or the plans as necessary.

		Description	Number
1.	⊖	_____	_____
2.	⬤	_____	_____
3.	s	_____	_____
4.	S$_{RC_a}$	_____	_____

40

5. 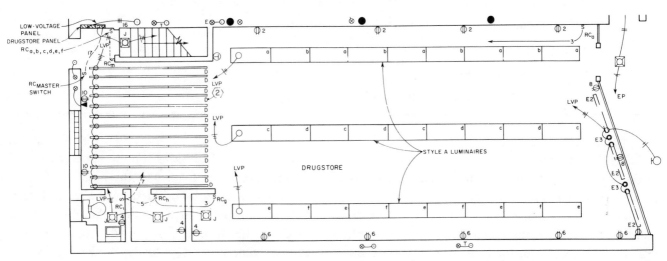 _____ _____

6. _____ _____

7. _____ _____

8. The elevation of the finished first floor is _____ feet.

9. The drugstore area has a(an) _____ ceiling.

10. The north wall of the drugstore is constructed of _____ and is _____ inches thick.

11. The south wall of the drugstore is constructed of _____ and _____ . This wall is _____ inches thick.

12. The ceiling height in the drugstore is _____ .

13. The concrete floor in the drugstore is _____ inches thick.

14. The outside dimensions of the building are _____ by _____ .

15. The basement area assigned to the drugstore is equal to _____ square feet.

NOTE: FOR COMPLETE BLUEPRINT, REFER TO BLUEPRINT E-2 IN BACK OF TEXT.

UNIT 4
Branch-circuit Requirements

OBJECTIVES

After completing the study of this unit, the student will be able to

- calculate the minimum allowable loading for a given area.
- determine the minimum number of branch circuits permitted.
- calculate the current in each line of a three-phase, four-wire system.
- determine the correct size for the branch-circuit protection.
- select the correct receptacles.
- derate conductors as required.
- calculate the voltage drop in single-phase circuits.

The electrician often must assume the responsibility for making decisions on the sizing, protection, and installation of branch circuits. Even on well-engineered projects, the electrician will be asked to make numerous choices based on economic considerations. In the commercial building in the plans, the selection of the wire type to be used and, therefore, the sizing of the conduit, the derating of the conductor, and the allowance for voltage drop are decisions to be made by the electrician.

BRANCH-CIRCUIT CALCULATION

According to *Article 100* of the *NEC,* a branch circuit is that portion of the wiring system between the final overcurrent device protecting the circuit and the outlet(s). Thus, all outlets are supplied by branch circuits. An outlet is defined in *Article 100* of the *NEC* as a point on the wiring system at which current is taken to supply equipment for a specific use. It can be seen that once the outlet requirements are tabulated, the number and sizes of the branch circuits can be determined.

Articles 210, 215 and *220* of the *NEC* govern the required calculations for the expected branch-circuit and feeder loads and the number of branch circuits necessary. To illustrate the procedure for determining the branch circuits, the branch-circuit loads in the drugstore are divided into four categories: general lighting loads, show window lighting, motor load, and other loads. The Code requirements for load calculations are summarized as follows:

- use *Table 220-2(b)* to find the value of watts per square feet (0.093 m²) for general lighting loads or use the actual wattage or volt-ampere ratings of fixtures, whichever is larger.

- for show window lighting, allow 200 watts per linear foot (305 mm) or use the actual ratings of the fixtures, whichever is larger, *Section 220-2(c), Exception No. 3.*

- install at least one receptacle outlet above each 12-foot (3.66 m) section of show window, *Section 210-62.*

- the continuous loading of branch circuits and feeders shall not exceed 80% of the ratings for same, *Sections 210-22(c)* and *220-2(a)*.

- allow 600 volt-amperes for heavy-duty lampholders, *Section 220-2(c)*.

- allow 180 volt-amperes for each receptacle outlet, *Section 220-2(c)*.

- for specific loads, use the actual ratings, *Section 220-2(c)*.

- for motors, see *Article 430*.

- for transformers, see *Article 450*.

- for electric heating, see *Article 424*.

- for appliances, see *Article 422*.

- for air conditioning and refrigeration, see *Article 440*.

- for multioutlet assemblies, allow 180 volt-amperes for each foot (305 mm) of the assembly of 180 volt-amperes for each 5-foot (1.52 m) section, depending upon how heavily the assembly will be loaded, *Section 220-2(c), Exception No. 1*.

General Lighting Loads

Lighting for Display Area of Store

The minimum lighting load to be included in the calculations for a given type of occupancy is determined from *Table 220-2(b)*. For a "store"

Table 220-2(b).   General Lighting Loads by Occupancies

Type of Occupancy	Unit Load per Sq. Ft. (Volt-Amperes)
Armories and Auditoriums	1
Banks	3½**
Barber Shops and Beauty Parlors	3
Churches	1
Clubs	2
Court Rooms	2
*Dwelling Units	3
Garages — Commercial (storage)	½
Hospitals	2
*Hotels and Motels, including apartment houses without provisions for cooking by tenants	2
Industrial Commercial (Loft) Buildings	2
Lodge Rooms	1½
Office Buildings	3½**
Restaurants	2
Schools	3
Stores	3
Warehouses (storage)	¼
In any of the above occupancies except one-family dwellings and individual dwelling units of two-family and multifamily dwellings: Assembly Halls and Auditoriums Halls, Corridors, Closets, Stairways Storage Spaces	1 ½ ¼

For SI units: one square foot = 0.093 square meter.

* All receptacle outlets of 20-ampere or less rating in one-family, two-family and multifamily dwellings and in guest rooms of hotels and motels [except those connected to the receptacle circuits specified in Section 220-3(b)] shall be considered as outlets for general illumination, and no additional load calculations shall be required for such outlets.

** In addition a unit load of 1 volt-ampere per square foot shall be included for general purpose receptacle outlets when the actual number of general purpose receptacle outlets is unknown.

occupancy, the table indicates that the unit load per square foot is 3 watts. Therefore the calculated lighting load in watts for the drugstore is: 60 ft. × 23.25 ft. × 3 watts/sq. ft. = 4,185 watts. Since this wattage value is the minimum allowance, the actual connected load should be found and the larger of the two values used to determine the load allowance. The total connected general lighting load is:

27 Style A luminaires		
@ 200 volt-amperes	5,400 volt-amperes	
4 Style J luminaires		
@ 150 watts	600 watts	
39 Style D luminaires		
@ 100 volt-amperes	3,900 volt-amperes	
	9,900 watts	

An interesting situation exists at this point. The connected load consists partly of lighting units having ballasts. A later unit on lamps will show that a fluorescent fixture, which is similar to those used for the main lighting of the drugstore, uses four 40-watt lamps (160 watts), but has an ampere rating of 1.6 amperes at 120 volts or 192 volt-amperes. In actual practice, this rating is rounded off to 200 *volt-amperes*. This value is then added to the *wattage* values of the incandescent lamps to determine the total load.

When it is possible to do so, the authors of this text will discriminate between volt-ampere and watt ratings. In all cases, however, the terminology of the *NEC* will be used. *Whenever branch circuits supply lighting loads having ballasts or transformers, the load calculations must be based on the total volt-ampere ratings and not on the total wattage ratings of the lamps.*

Storage Area Lighting. According to *Table 220-2(b)*, the minimum lighting load for storage spaces is 1/4 watt per square foot. Thus, the computed minimum load allowance for the basement area is: 1,012 sq. ft. × 1/4 watt per sq. ft. = 253 watts. The actual connected load is the total of 9 style L luminaires @ 100 volt-amperes each or 900 volt-amperes. Using the larger values for both the display area and storage area lighting, the general lighting load allowance is 9,900 watts + 900 volt-amperes = 10,800 volt-amperes.

Display Window Lighting

The allowance for show window lighting is based on a minimum of 200 watts per linear foot (305 mm) of window. Since the drugstore has 16 feet (4.88 m) of show window, the lighting allowance is:

16 ft (4.88 m) × 200 watts per ft (305 mm)
= 3,200 watts)

The actual connected load consists of luminaires and an estimated receptacle load:

3 Style E2 luminaires		
@ 400 watts	= 1,200 watts	
2 Style E3 luminaires		
@ 150 watts	= 300 watts	
3 receptacle outlets		
@500 watts	= 1,500 watts	
TOTAL CONNECTED LOAD	3,000 watts	

As the connected load is less than the calculated allowance of 3,200 watts, this minimum allowance (3,200 watts) will be used for the show window lighting load.

Motor Load

The drugstore is provided with air-conditioning equipment. The data on the nameplate of the air-conditioning unit is as follows:

Voltage:	208 volts, three phase three wire, 60 hertz
Compressor motor:	FLA 20.2, LRA 90, three phase
Condenser motor:	FLA 3.2, 1/4 hp, single phase
Evaporator motor:	FLA 3.2, 1/4 hp, single phase
Minimum ampacity of supply conductors:	40 amperes
*Maximum fuse size:	40 amperes

*It is important to note that when the nameplate specifies *fuses*, the equipment is intended to be protected by fuses only. Circuit breakers shall not be used. When the nameplate specifies *overcurrent protection*, either fuses or circuit breakers may be used.

The electrician must install supply conductors and overcurrent protection (fuses) in accordance with the nameplate data. Tests conducted by the manufacturer and Underwriters Laboratories determine the proper conductor ampacity and overcurrent protection ratings required.

Another method for calculating the circuit requirements for this air conditioner is shown in the following example. (This method, however, is more complicated than taking the data from the nameplate of the unit.)

20.2 amperes $\times$ 208 volts $\times$ 1.73	=	7,272 volt-amperes
3.2 amperes $\times$ 208 volts	=	665 volt-amperes
3.2 amperes $\times$ 208 volts	=	665 volt-amperes
Total motor load		8,602 volt-amperes
Plus 25% of largest motor (20.2 amperes $\times$ 208 volts $\times$ 1.73 $\times$ 0.25)	=	1,818 volt-amperes
Total calculated load		10,420 volt-amperes

The Code requirements for air-conditioning equipment are detailed in *NEC Article 440*.

Three-phase current calculations are discussed later in this unit. See Unit 11 for a complete discussion on air-conditioning equipment, the theory of operation, and installation methods and calculations.

Other Loads, *NEC Section 220-2(c)*

Receptacle Outlets. In addition to the three convenience receptacle outlets for the display window, 17 other receptacle outlets must be installed. According to *Section 220-2(c)*, the load allowance for these outlets is 180 volt-amperes per outlet. Therefore, the power capacity required will be:

180 volt-amperes per outlet $\times$ 17 outlets
= 3,060 volt-amperes

Although the actual load to be connected to these receptacles is not known, the load is assumed to be the minimum required by the *NEC*.

Sign Outlet. The actual load requirement for the sign is 100 volt-amperes. Since this is a specific load, the actual ampere rating of the load will be used in the calculations, (see *NEC Section 220-3*).

Loading Schedule

Table 4-1 summarizes the branch-circuit loading as determined in this unit for the drugstore occupancy in the commercial building.

BRANCH CIRCUITS

The *minimum* number of branch circuits required for the drugstore is determined from the loading schedule. Other factors such as the switching arrangement and the convenience of installation will be important considerations in determining the *actual* number of branch circuits to be installed.

The *NEC*, in *Section 220-3*, establishes certain minimum requirements that must be followed when determining the branch circuits. However, the electrical system designer has the option of exceeding the minimum requirements set by the *NEC*. The *NEC* standards are set to provide a reasonable flexibility in the design to meet any future requirements.

Throughout the design of the electrical system for the commercial building, the following standards are followed. These standards are either included in the specifications or shown as alternatives in the *National Electrical Code*.

1. The allowable load current of each conductor shall be reduced in accordance with *Note 8* of *Tables 310-16* through *310-19*.

2. Under normal conditions, any load may operate for more than three continuous hours. Thus, the loading of an overcurrent device located in a panelboard shall not exceed 80% of its rating, per *NEC Section 384-16(c)*.

3. The minimum size for a branch-circuit conductor shall be No. 12 AWG.

The arrangement of the drugstore branch circuits is indicated in the appropriate panel schedule.

Lighting and Appliance Branch Circuits

Branch-circuit overcurrent devices in the commercial building are sized according to the ampacity of the conductors as permitted by *NEC Section 240-3* and *Table 310-16* for copper conductors in raceway.

The specifications direct the electrician to install copper conductors of a size which is not smaller than No. 12 AWG. The conductors may have type TW, THW, THWN, or XHHW insulation, all of which have an ampacity of 20 amperes.

Therefore, the minimum number of circuits can be determined by dividing the lighting and receptacle outlet load by the maximum permitted load for a 20-ampere branch circuit.

Maximum permitted load per branch circuit:

20 amperes $\times$ 120 volts $\times$ 0.8

= 1,920 volt-amperes

Lighting and receptacle load:

27,580 VA − 10,420 VA

= 17,160 volt-amperes

Therefore, the minimum number of branch circuits is found by division, as follows:

17,160 VA ÷ 1,920 VA/circuit

= 9 circuits

A check of the panel schedule (found in the specifications) shows that 13 circuits are used for the drugstore load. To accommodate any future growth in the electrical requirements, eight spare circuits are provided.

Three-phase Current Calculations

It may be necessary to calculate the total current to find the size of the short-circuit protection required for a multimotor air-conditioning unit. Since the currents in a three-phase system are vectors, the rated load currents must be summed by vector addition to determine the actual currents

TABLE 4-1 DRUGSTORE LOADING SCHEDULE

Loading	Applicable NEC Section	Load, in Watts		
		NEC	Actual	Design
GENERAL LIGHTING				
Store Area:				
1,395 sq. ft. $\times$ 3 watts/sq. ft.	220-2	4,185		
27 Style A luminaires @ 200 VA			5,400	
4 Style J luminaires @ 150 watts			600	
39 Style D luminaires @ 100 VA			3,900	
Storage Area:				
1,012 sq. ft. $\times$ 1/4 watts/sq. ft.	220-2	253		
9 Style L luminaires @ 100 VA			900	
Total general lighting		4,438	10,800	
Value to be used				10,800
SHOW WINDOW				
16 ft. $\times$ 200 watts/ft.	220-12	3,200		
3 Style E2 luminaires @ 400 watts			1,200	
2 Style E3 luminaires @ 150 watts			300	
3 receptacle outlets @ 500 watts			1,500	
Total show window		3,200	3,000	
Value to be used				3,200
MOTOR LOAD				
CM 20.2 amperes $\times$ 1.73 $\times$ 208 volts	440-32	7,272	7,272	
EFM 3.2 amperes $\times$ 208 volts		665	665	
CFM 3.2 amperes $\times$ 208 volts		665	665	
25% allowance for CM motor		1,818		
Total motor load		10,420	8,602	
Value to be used				10,420
OTHER				
Receptacle outlets:				
17 outlets $\times$ 180 VA/outlet	220-2	3,060		
Sign outlet:				
1 outlet @ 100 VA	220-2	100		
Value to be used		3,160		3,160
TOTAL LOAD				27,580 watts

flowing in a given conductor. The actual current value depends upon several variables including the phase connection sequence (ABC or ACB), the power factor of each load, and the difference in the rated currents. For simplicity, the arithmetic sum of the currents is generally used to calculate the individual line currents. This sum replaces the lengthy calculations required to determine the actual currents. However, the student should realize that a meter reading of the amperage taken while the unit is operating will not be the same as the calculated current values. For example, figure 4-1 shows a circuit where an ammeter will read 20.2 amperes in phase A and slightly less than 26 amperes in phases B and C.

The arithmetic sum of the current in phases B and C is:

$$20.2 + 3.2 + 3.2 = 26.6 \text{ amperes}$$

The minimum circuit ampacity is found by adding 25% of the motor-compressor rating in accordance with *NEC Sections 440-33, 440-35* and *430-24.*

$$26.6 + (20.2 \times 0.25) = 31.65 \text{ amperes}$$

Referring to *NEC Table 310-16,* page 53, it can be seen that a No. 8 TW conductor will meet this requirement. The branch-circuit short-circuit protection can now be calculated. The value used in the calculation is the motor compressor rated load current multiplied by 1.75 (from *NEC Table 430-152* for a time-delay fuse protective device). Also refer to *Sections 440-22(b), 430-52,* and *430-53(c).*

$$20.2 \text{ amperes} \times 1.75 = 35.35 \text{ amperes}$$

Therefore, a 40-ampere fuse is used. If the motor does not start with this device, the short-circuit protection may be increased but shall not exceed 20.2 amperes $\times$ 2.25 = 45.45 amperes (45-ampere time-delay, dual element fuse). This increase is permitted by *Section 240-3, Exception 3,* which refers the reader to *Article 422 (Appliances), Article 430 (Motors),* and *Article 440 (Air-conditioning and refrigeration equipment).*

SPECIAL CONSIDERATIONS

Following the selection of the circuit conductors and the branch-circuit protection, there are a number of special considerations that generally are factors for each installation. The electrician is usually given the responsibility of planning the

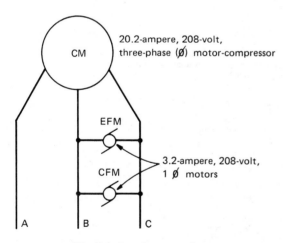

CM — 20.2-ampere, 208-volt, three-phase (∅) motor-compressor

EFM

CFM — 3.2-ampere, 208-volt, 1 ∅ motors

A   B   C

Fig. 4-1 Load connections.

Notes to Tables

1. Tables 3A, 3B and 3C apply only to complete conduit or tubing systems and are not intended to apply to short sections of conduit or tubing used to protect exposed wiring from physical damage.

2. Equipment grounding conductors, when installed, shall be included when calculating conduit or tubing fill. The actual dimensions of the equipment grounding conductor (insulated or bare) shall be used in the calculation.

3. When conduit nipples having a maximum length not to exceed 24 inches (610 mm) are installed between boxes, cabinets, and similar enclosures, the nipple shall be permitted to be filled to 60 percent of its total cross-sectional area, and Note 8 of Tables 310-16 through 310-19 does not apply to this condition.

4. For conductors not included in Chapter 9, such as compact or multiconductor cables, the actual dimensions shall be used.

5. See Table 1 for allowable percentage of conduit or tubing fill.

Table 1 is based on common conditions of proper cabling and alignment of conductors where the length of the pull and the number of bends are within reasonable limits. It should be recognized that for certain conditions a larger size conduit or a lesser conduit fill should be considered.

Table 1. Percent of Cross Section of Conduit and Tubing for Conductors

(See Table 2 for Fixture Wires)

Number of Conductors	1	2	3	4	Over 4
All conductor types except lead-covered (new or rewiring)	53	31	40	40	40
Lead-covered conductors	55	30	40	38	35

Note 1. See Tables 3A, 3B and 3C for number of conductors all of the same size in trade sizes of conduit ½ inch through 6 inch.

Note 2. For conductors larger than 750 MCM or for combinations of conductors of different sizes, use Tables 4 through 8, Chapter 9, for dimensions of conductors, conduit and tubing.

Note 3. Where the calculated number of conductors, all of the same size, includes a decimal fraction, the next higher whole number shall be used where this decimal is 0.8 or larger.

Note 4. When bare conductors are permitted by other sections of this Code, the dimensions for bare conductors in Table 8 of Chapter 9 shall be permitted.

Note 5. A multiconductor cable of two or more conductors shall be treated as a single conductor cable for calculating percentage conduit fill area. For cables that have elliptical cross section, the cross-sectional area calculation shall be based on using the major diameter of the ellipse as a circle diameter.

Table 2.   Maximum Number of Fixture Wires in Trade Sizes of Conduit or Tubing

(40 Percent Fill Based on Individual Diameters)

Conduit Trade Size (Inches)	½					¾					1					1¼					1½					2				
Wire Types	18	16	14	12	10	18	16	14	12	10	18	16	14	12	10	18	16	14	12	10	18	16	14	12	10	18	16	14	12	10
PTF, PTFF, PGFF, PGF, PFF, PF, PAF, PAFF, ZF, ZFF	23	18	14			40	31	24			65	50	39			115	90	70			157	122	95			257	200	156		
TFFN, TFN	19	15				34	26				55	43				97	76				132	104				216	169			
SF-1	16					29					47					83					114					186				
SFF-1, FFH-1	15					26					43					76					104					169				
CF	13	10	8	4	3	23	18	14	7	6	38	30	23	12	9	66	53	40	21	16	91	72	55	29	22	149	118	90	48	37
TF	11	10				20	18				32	30				57	53				79	72				129	118			
RFH-1	11					20					32					57					79					129				
TFF	11	10				20	17				32	27				56	49				77	66				126	109			
AF	11	9	7	4	3	19	16	12	7	5	31	26	20	11	8	55	46	36	19	15	75	63	49	27	20	123	104	81	44	34
SFF-2	9	7	6			16	12	10			27	20	17			47	36	30			65	49	42			106	81	68		
SF-2	9	8	6			16	14	11			27	23	18			47	40	32			65	55	43			106	90	71		
FFH-2	9	7				15	12				25	19				44	34				60	46				99	75			
RFH-2	7	5				12	10				20	16				36	28				49	38				80	62			
KF-1, KFF-1, KF-2, KFF-2	36	32	22	14	9	64	55	39	25	17	103	89	63	41	28	182	158	111	73	49	248	216	152	100	67	406	353	248	163	110

Table 3A.   Maximum Number of Conductors in Trade Sizes of Conduit or Tubing

(Based on Table 1, Chapter 9)

Type Letters	Conductor Size AWG, MCM	½	¾	1	1¼	1½	2	2½	3	3½	4	4½	5	6
TW, T, RUH, RUW, XHHW (14 thru 8)	14	9	15	25	44	60	99	142	171					
	12	7	12	19	35	47	78	111	131	176				
	10	5	9	15	26	36	60	85						
	8	2	4	7	12	17	28	40	62	84	108			
RHW and RHH (without outer covering), THW	14	6	10	16	29	40	65	93	143	192				
	12	4	8	13	24	32	53	76	117	157				
	10	4	6	11	19	26	43	61	95	127	163			
	8	1	3	5	10	13	22	32	49	66	85	106	133	
TW, T, THW, RUH (6 thru 2), RUW (6 thru 2), FEPB (6 thru 2), RHW and RHH (without outer covering)	6	1	2	4	7	10	16	23	36	48	62	78	97	141
	4	1	1	3	5	7	12	17	27	36	47	58	73	106
	3	1	1	2	4	6	10	15	23	31	40	50	63	91
	2	1	1	2	4	5	9	13	20	27	34	43	54	78
	1		1	1	3	4	6	9	14	19	25	31	39	57
	0		1	1	2	3	5	8	12	16	21	27	33	49
	00			1	1	3	5	7	10	14	18	23	29	41
	000			1	1	2	4	6	9	12	15	19	24	35
	0000				1	1	3	5	7	10	13	16	20	29
	250			1	1	1	2	4	6	8	10	13	16	23
	300			1	1	1	2	3	5	7	9	11	14	20
	350				1	1	1	3	4	6	8	10	12	18
	400				1	1	1	2	4	5	7	9	11	16
	500				1	1	1	1	3	4	6	7	9	14
	600					1	1	1	3	4	5	6	7	11
	700					1	1	1	2	3	4	5	7	10
	750					1	1	1	2	3	4	5	6	9

Table 3B. Maximum Number of Conductors in Trade Sizes of Conduit or Tubing

(Based on Table 1, Chapter 9)

Type Letters	Conductor Size AWG, MCM	½	¾	1	1¼	1½	2	2½	3	3½	4	4½	5	6
THWN,	14	13	24	39	69	94	154							
	12	10	18	29	51	70	114	164						
	10	6	11	18	32	44	73	104	160					
	8	3	5	9	16	22	36	51	79	106	136			
THHN,	6	1	4	6	11	15	26	37	57	76	98	125	154	
FEP (14 thru 2),	4	1	2	4	7	9	16	22	35	47	60	75	94	137
FEPB (14 thru 8),	3	1	1	3	6	8	13	19	29	39	51	64	80	116
PFA (14 thru 4/0)	2	1	1	3	5	7	11	16	25	33	43	54	67	97
PFAH (14 thru 4/0)	1	1	1	3	5	8	12	18	25	32	40	50	72	
Z (14 thru 4/0)	0		1	1	3	4	7	10	15	21	27	33	42	61
XHHW (4 thru 500MCM)	00		1	1	2	3	6	8	13	17	22	28	35	51
	000		1	1	1	3	5	7	11	14	18	23	29	42
	0000		1	1	1	2	4	6	9	12	15	19	24	35
	250			1	1	1	3	4	7	10	12	16	20	28
	300			1	1	1	3	4	6	8	11	13	17	24
	350			1	1	1	2	3	5	7	9	12	15	21
	400				1	1	1	3	5	6	8	10	13	19
	500				1	1	1	2	4	5	7	9	11	16
	600				1	1	1	1	3	4	5	7	9	13
	700				1	1	1	1	3	4	5	6	8	11
	750					1	1	1	2	3	4	6	7	11
XHHW	6	1	3	5	9	13	21	30	47	63	81	102	128	185
	600				1	1	1	1	3	4	5	7	9	13
	700					1	1	1	3	4	5	6	8	11
	750					1	1	1	2	3	4	6	7	10

Table 3C. Maximum Number of Conductors in Trade Sizes of Conduit or Tubing

(Based on Table 1, Chapter 9)

Type Letters	Conductor Size AWG, MCM	½	¾	1	1¼	1½	2	2½	3	3½	4	4½	5	6
RHW,	14	3	6	10	18	25	41	58	90	121	155			
	12	3	5	9	15	21	35	50	77	103	132			
	10	2	4	7	13	18	29	41	64	86	110	138		
	8	1	2	4	7	9	16	22	35	47	60	75	94	137
RHH (with outer covering)	6	1	1	2	5	6	11	15	24	32	41	51	64	93
	4	1	1	1	3	5	8	12	18	24	31	39	50	72
	3	1	1	1	3	4	7	10	16	22	28	35	44	63
	2	1	1	1	3	4	6	9	14	19	24	31	38	56
	1	1	1	1	1	3	5	7	11	14	18	23	29	42
	0		1	1	1	2	4	6	9	12	16	20	25	37
	00			1	1	1	3	5	8	11	14	18	22	32
	000			1	1	1	3	4	7	9	12	15	19	28
	0000			1	1	1	2	4	6	8	10	13	16	24
	250				1	1	1	3	5	6	8	11	13	19
	300				1	1	1	3	4	5	7	9	11	17
	350				1	1	1	2	4	5	6	8	10	15
	400					1	1	1	3	4	6	7	9	14
	500				1	1	1	1	3	4	5	6	8	11
	600						1	1	1	3	4	5	6	9
	700						1	1	1	3	3	4	6	8
	750						1	1	1	3	3	4	5	8

Table 4.   Dimensions and Percent Area of Conduit and of Tubing

Areas of Conduit or Tubing for the Combinations of Wires Permitted in Table 1, Chapter 9.

| Trade Size | Internal Diameter Inches | Area — Square Inches | | | | | | | | |
| | | Total 100% | Not Lead Covered | | | Lead Covered | | | | |
			2 Cond. 31%	Over 2 Cond. 40%	1 Cond. 53%	1 Cond. 55%	2 Cond. 30%	3 Cond. 40%	4 Cond. 38%	Over 4 Cond. 35%
½	.622	.30	.09	.12	.16	.17	.09	.12	.11	.11
¾	.824	.53	.16	.21	.28	.29	.16	.21	.20	.19
1	1.049	.86	.27	.34	.46	.47	.26	.34	.33	.30
1¼	1.380	1.50	.47	.60	.80	.83	.45	.60	.57	.53
1½	1.610	2.04	.63	.82	1.08	1.12	.61	.82	.78	.71
2	2.067	3.36	1.04	1.34	1.78	1.85	1.01	1.34	1.28	1.18
2½	2.469	4.79	1.48	1.92	2.54	2.63	1.44	1.92	1.82	1.68
3	3.068	7.38	2.29	2.95	3.91	4.06	2.21	2.95	2.80	2.58
3½	3.548	9.90	3.07	3.96	5.25	5.44	2.97	3.96	3.76	3.47
4	4.026	12.72	3.94	5.09	6.74	7.00	3.82	5.09	4.83	4.45
4½	4.506	15.94	4.94	6.38	8.45	8.77	4.78	6.38	6.06	5.56
5	5.047	20.00	6.20	8.00	10.60	11.00	6.00	8.00	7.60	7.00
6	6.065	28.89	8.96	11.56	15.31	15.89	8.67	11.56	10.98	10.11

Table 5.   Dimensions of Rubber-Covered and Thermoplastic-Covered Conductors

| Size AWG MCM | Types RFH-2, RH, RHH,*** RHW,*** SF-2 | | | | Types TF, T, THW,† TW, RUH,** RUW** | | Types TFN, THHN, THWN | | Types**** FEP, FEPB, FEPW, TFE, PF, PFA, PFAH, PGF, PTF, Z, ZF, ZFF | | | | Type XHHW, ZW†† | | Types KF-1, KF-2, KFF-1, KFF-2 | |
| | Approx. Diam. Inches | | Approx. Area Sq. In. | | Approx. Diam. Inches | Approx. Area Sq. In. | Approx. Diam. Inches | Approx. Area Sq. In. | Approx. Diam. Inches | | Approx. Area Sq. Inches | | Approx. Diam. Inches | Approx. Area Sq. In. | Approx. Diam. Inches | Approx. Area Sq. In. |
Col. 1	Col. 2		Col. 3		Col. 4	Col. 5	Col. 6	Col. 7	Col. 8		Col. 9		Col. 10	Col. 11	Col. 12	Col. 13
18	.146		.0167		.106	.0088	.089	.0062	.081		.0052		...	....	.065	.0033
16	.158		.0196		.118	.0109	.100	.0079	.092		.0066		...	....	.070	.0038
14	30 mils	.171	.0230		.131	.0135	.105	.0087	.105	.105	.0087	.0087	...	....	.083	.0054
14	45 mils	.204*	.0327*		...	....	...	....	...	...	....	....	.129	.0131	...	....
14					.162†	.0206†	...	....	...	...	....	....			.102	.0082
12	30 mils	.188	.0278		.148	.0172	.122	.0117	.121	.121	.0115	.0115	...	....		
12	45 mils	.221*	.0384*		...	....	...	....	...	...	....	....	.146	.0167		
12					.179†	.0252†	...	....	...	...	....	....			.124	.0121
10			.242	.0460	.168	.0222	.153	.0184	.142	.142	.0158	.0158	.166	.0216		
10					.199†	.0311†	...	....	...	...	....	....				
8			.328	.0845	.245	.0471	.218	.0373	.206	.186	.0333	.0272	.241	.0456		
8					.276†	.0598†	...	....	...	...	....	....				
6	.397		.1238		.323	.0819	.257	.0519	.244	.302	.0468	.0716	.282	.0625	...	....
4	.452		.1605		.372	.1087	.328	.0845	.292	.350	.0670	.0962	.328	.0845	...	....
3	.481		.1817		.401	.1263	.356	.0995	.320	.378	.0804	.1122	.356	.0995	...	....
2	.513		.2067		.433	.1473	.388	.1182	.352	.410	.0973	.1320	.388	.1182	...	....
1	.588		.2715		.508	.2027	.450	.1590	.420	...	.1385	....	.450	.1590	...	....
0	.629		.3107		.549	.2367	.491	.1893	.462	...	.1676	....	.491	.1893	...	....
00	.675		.3578		.595	.2781	.537	.2265	.498	...	.1948	....	.537	.2265	...	....
000	.727		.4151		.647	.3288	.588	.2715	.560	...	.2463	....	.588	.2715	...	....
0000	.785		.4840		.705	.3904	.646	.3278	.618	...	.3000	....	.646	.3278	...	....

Table 5 (Continued)

Size AWG MCM	Types RFH-2, RH, RHH,*** RHW,*** SF-2		Types TF, T, T4W,† TW, RUH,** RUW**		Types TFN, THHN, THWN		Types**** FEP, FEPB, FEPW, TFE, PF, PFA, PFAH, PGF, PTF, Z, ZF, ZFF		Type XHHW, ZW††	
	Approx. Diam. Inches	Approx. Area Sq. In.	Approx. Diam. Inches	Approx. Area Sq. In.	Approx. Diam. Inches	Approx. Area Sq. In.	Approx. Diam. Inches	Approx. Area Sq. Inches	Approx. Diam. Inches	Approx. Area Sq. In.
Col. 1	Col. 2	Col. 3	Col. 4	Col. 5	Col. 6	Col. 7	Col. 8	Col. 9	Col. 10	Col. 11
250	.868	.5917	.788	.4877	.716	.4026	. . .	. . . .	.716	.4026
300	.933	.6837	.843	.5581	.771	.4669	. . .	. . . .	.771	.4669
350	.985	.7620	.895	.6291	.822	.5307	. . .	. . . .	.822	.5307
400	1.032	.8365	.942	.6969	.869	.5931	. . .	. . . .	.869	.5931
500	1.119	.9834	1.029	.8316	.955	.7163	. . .	. . . .	.955	.7163
600	1.233	1.1940	1.143	1.0261	1.058	.8791	. . .	. . . .	1.073	.9043
700	1.304	1.3355	1.214	1.1575	1.129	1.0011	. . .	. . . .	1.145	1.0297
750	1.339	1.4082	1.249	1.2252	1.163	1.0623	. . .	. . . .	1.180	1.0936
800	1.372	1.4784	1.282	1.2908	1.196	1.1234	. . .	. . . .	1.210	1.1499
900	1.435	1.6173	1.345	1.4208	1.259	1.2449	. . .	. . . .	1.270	1.2668
1000	1.494	1.7530	1.404	1.5482	1.317	1.3623	. . .	. . . .	1.330	1.3893
1250	1.676	2.2062	1.577	1.9532	. . .	. . . .	. . .	. . . .	1.500	1.7671
1500	1.801	2.5475	1.702	2.2751	. . .	. . . .	. . .	. . . .	1.620	2.0612
1750	1.916	2.8832	1.817	2.5930	. . .	. . . .	. . .	. . . .	1.740	2.3779
2000	2.021	3.2079	1.922	2.9013	. . .	. . . .	. . .	. . . .	1.840	2.6590

* The dimensions of Types RHH and RHW.
** No. 14 to No. 2.
† Dimensions of THW in sizes No. 14 to No. 8. No. 6 THW and larger is the same dimension as T.
*** Dimensions of RHH and RHW without outer covering are the same as THW No. 18 to No. 10, solid; No. 8 and larger, stranded.
**** In Columns 8 and 9 the values shown for sizes No. 1 thru 0000 are for TFE and Z only. The right-hand values in Columns 8 and 9 are for FEPB, Z, ZF, and ZFF only.
†† No. 14 to No. 2.

routing of the conduit to insure that the outlets are connected properly. As a result, the electrician must determine the length and the number of conductors in each raceway. In addition, the electrician must derate the conductors, make allowances for voltage drops, recognize the various receptacle types, and be able to install these receptacles correctly on the system.

Conduit Sizing

The required size of the conduit depends upon three factors: (1) the number of conductors to be installed, (2) the cross-sectional area of the conductor, and (3) the permissible conduit fill. The relationship of these factors is shown in *Tables 1* through *5* of *Chapter 9* of the *NEC*.

Once a determination is made as to the number of conductors to be installed to a certain point, the electrician can use the following procedure to find the conduit size.

- If all the conductors have the same insulation, the conduit size can be determined directly from *Tables 3A, 3B,* or *3C* of the *NEC*. For example, three No. 8 TW conductors can be installed to the air-conditioning unit. According to *NEC Table 3A*, for insulation type "TW" and conductor size "8," it is evident that 3/4-inch conduit is required for the three conductors.

To continue the example, assume that a No. 10 THWN equipment grounding conductor is to be installed with the three No. 8 conductors. (An installation of this type will meet the requirements of *Article 250, Grounding*, if a flexible conduit is used in the connection to the air-conditioning unit.) The No. 10 grounding conductor must be included in the calculation of the conduit fill according to *Chapter 9, Note 2* of the *NEC*.

- Refer to *Table 5* to find the area of the individual conductors. Calculate the total conductor area in square inches.

1 No. 10 THWN conductor	= 0.0184 sq. in.
3 No. 8 TW conductors @ 0.0471 sq. in. each	= 1.1413 sq. in.
Total conductor area	0.1597 sq. in.

- Refer to *Table 4* under the column "Not Lead Covered," and the subcolumn "Over 2

Cond. 40%" to find the next larger value of area (0.21 sq. in.). Then move to the column on the extreme left to find the conduit size. In this case, 3/4-inch conduit is required. Note that the actual internal area of 1/2-inch conduit is 0.30 sq. in. and the internal area of 3/4-inch conduit is 0.53 sq. in.

DETERMINE WIRE SIZE AND TYPE

Just as an electric motor can burn out, so can the insulation on a conductor become damaged as a result of extremely high temperatures. The insulation can soften, melt, and finally break down, causing grounds, shorts, and possible equipment damage and personal injury.

The source of this damaging heat comes from the surrounding room temperature (ambient temperature), and the heat generated by the current flowing through the wire. This heat can be calculated by:

$$\text{Watts} = I^2 R$$

Thus, the more current-carrying conductors that are installed in a single raceway or cable, the more damaging is the accumulative effect of the heat generated by the conductors. This heat must be kept to a safe, acceptable level by (1) restricting the number of conductors in a raceway or cable, or (2) limiting the amount of current that a given size conductor may carry. This means that various derating factors and correction factors must be applied to the ampacity of the conductor being installed. *Ampacity* is defined as "the current in amperes that a conductor can carry continuously under the condition of use without exceeding its temperature rating."

When called upon to connect an electrical load such as lighting, motors, heating, refrigeration, or air-conditioning equipment, the electrician must have a working knowledge of how to select the proper size and type of conductors to be installed. Installing the right size wire will assure that the voltage at the terminals of the equipment is within the minimums as set forth by the Code. It also assures that the insulation will be able to withstand the temperature conditions that exist, and that the conduit will be the correct size for ease of pulling the conductors into the conduit.

There are a number of things that must be considered when selecting wires. The Code is referred to continually. The Code contains terms such as *total load, computed load, intermittent load, noncontinuous load, connected load, load current rating, maximum allowable load current, rated load current, branch-circuit selection current, derating factors, correction factors, ambient temperature, ampacity,* and so on. These terms have a specific meaning and are very important in the selection of conductors. The electrician should become familiar with these terms.

One of the first steps in understanding conductors is to refer to *Article 310* of the *NEC* entitled Conductors for General Wiring. This article contains such topics as insulation types, conductors in parallel, wet and dry locations, marking, maximum operating temperatures, permitted use, trade name, direct burial, ampacity tables, derating factors, and correction factors. It also contains many other important issues that must be considered.

Tables 310-16 through *310-19* are referred to continually by electricians, design engineers, and electrical inspectors. These tables show the allowable ampacities of insulated conductors and are easy to use. A person merely looks for the needed information. The temperature limitations, the types of insulation, the material the conductor is made of (copper, copper-clad aluminum, aluminum), the conductor size in AWG (American Wire Gauge) or in CM (circular mils) or in MCM (thousand circular mils), the ampacity of the conductor, and correction factors for high temperature are all indicated in the table.

Table 310-16 is the table used most since it is the table referred to when the conductors are installed in a raceway. The other tables can be found by referring to the *National Electrical Code.*

Table 310-16. Ampacities of Insulated Conductors Rated 0-2000 Volts, 60° to 90°C

Not More Than Three Conductors in Raceway or Cable or Earth (Directly Buried), Based on Ambient Temperature of 30°C (86°F)

Size	Temperature Rating of Conductor, See Table 310-13								Size
	60°C (140°F)	75°C (167°F)	85°C (185°F)	90°C (194°F)	60°C (140°F)	75°C (167°F)	85°C (185°F)	90°C (194°F)	
AWG MCM	TYPES †RUW, †T, †TW, †UF	TYPES †FEPW, †RH, †RHW, †RUH, †THW, †THWN, †XHHW, †USE, †ZW	TYPES V, MI	TYPES TA, TBS, SA, AVB, SIS, †FEP, †FEPB, †RHH †THHN, †XHHW*	TYPES †RUW, †T, †TW, †UF	TYPES †RH, †RHW, †RUH, †THW †THWN, †XHHW, †USE	TYPES V, MI	TYPES TA, TBS, SA, AVB, SIS, †RHH, †THHN, †XHHW*	AWG MCM
	COPPER				ALUMINUM OR COPPER-CLAD ALUMINUM				
18	....	....	....	14	....	....	....	....	
16	....	....	18	18	....	....	....	....	
14	20†	20†	25	25†	....	....	....	....	
12	25†	25†	30	30†	20†	20†	25	25†	12
10	30	35†	40	40†	25	30†	30	35†	10
8	40	50	55	55	30	40	40	45	8
6	55	65	70	75	40	50	55	60	6
4	70	85	95	95	55	65	75	75	4
3	85	100	110	110	65	75	85	85	3
2	95	115	125	130	75	90	100	100	2
1	110	130	145	150	85	100	110	115	1
0	125	150	165	170	100	120	130	135	0
00	145	175	190	195	115	135	145	150	00
000	165	200	215	225	130	155	170	175	000
0000	195	230	250	260	150	180	195	205	0000
250	215	255	275	290	170	205	220	230	250
300	240	285	310	320	190	230	250	255	300
350	260	310	340	350	210	250	270	280	350
400	280	335	365	380	225	270	295	305	400
500	320	380	415	430	260	310	335	350	500
600	355	420	460	475	285	340	370	385	600
700	385	460	500	520	310	375	405	420	700
750	400	475	515	535	320	385	420	435	750
800	410	490	535	555	330	395	430	450	800
900	435	520	565	585	355	425	465	480	900
1000	455	545	590	615	375	445	485	500	1000
1250	495	590	640	665	405	485	525	545	1250
1500	520	625	680	705	435	520	565	585	1500
1750	545	650	705	735	455	545	595	615	1750
2000	560	665	725	750	470	560	610	630	2000

AMPACITY CORRECTION FACTORS

Ambient Temp. °C	For ambient temperatures other than 30°C, multiply the ampacities shown above by the appropriate factor shown below.								Ambient Temp. °F
31-40	.82	.88	.90	.91	.82	.88	.90	.91	87-104
41-45	.71	.82	.85	.87	.71	.82	.85	.87	105-113
46-50	.58	.75	.80	.82	.58	.75	.80	.82	114-122
51-60	....	.58	.67	.71	....	.58	.67	.71	123-141
61-70	....	.35	.52	.58	....	.35	.52	.58	142-158
71-80	....	....	.30	.41	....	....	.30	.41	159-176

† The overcurrent protection for conductor types marked with an obelisk (†) shall not exceed 15 amperes for 14 AWG, 20 amperes for 12 AWG, and 30 amperes for 10 AWG copper; or 15 amperes for 12 AWG and 25 amperes for 10 AWG aluminum and copper-clad aluminum after any correction factors for ambient temperature and number of conductors have been applied.

* For dry locations only. See 75°C column for wet locations.

Reprinted with permission from NFPA 70-1984, *National Electrical Code®*, Copyright © 1983, National Fire Protection Association, Quincy, Massachusetts 02269. This reprinted material is not the complete and official position of the NFPA on the referenced subject, which is represented only by the standard in its entirety.

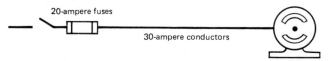

Fig. 4-2 Where a larger conductor is installed for any reason, such as to minimize voltage drop, the ampere rating of the overcurrent device determines the branch-circuits classification. In this example, 30-ampere conductors have been installed and connected to the 20-ampere overcurrent device. This circuit is classified as a 20-ampere circuit.

Table 310-16 gives the ampacities of the conductors before applying derating factors or correction factors. These are:

- The conductors must have correction factors applied when high temperatures are involved.
- The conductors must be derated when more than 3 current-carrying conductors are installed in a raceway or cable. See *Note 8, Table 310-16.*
- The conductors must be derated when supplying continuous loads or combinations of continuous and noncontinuous loads. Continuous load is defined as "a load where the maximum current is expected to continue for three hours or more."
- The conductors must be sized 25% greater than the full load current of a motor, i.e. FLA times 1.25 equals minimum ampacity of conductor.
- Conductors for air-conditioners (hermetically sealed motors) are sized according to the branch-circuit selection current. This information is found on the label of the equipment.
- Consideration must be given to voltage drop.

CORRECTION FACTORS FOR HIGH TEMPERATURES

The ampacities as shown in *Tables 310-16* through *310-19* must always be reduced according to the correction factors found at the bottom of these tables when *ambient* (surrounding) temperatures are above 30°C (86°F).

For example, consider that 3 current-carrying No. 3 RHW copper conductors are to be installed in one raceway in an ambient temperature of 35°C (95°F). The maximum permitted load current for these conductors is determined as follows:

The ampacity of No. 3 RHW copper conductors from *Table 310-16* is 100 amperes.

Apply correction factor for 35°C (95°F):
$$100 \times 0.88 = 88 \text{ amperes}$$

Therefore, 88 amperes is the maximum permitted load current for the example given.

DERATING FACTORS FOR MORE THAN 3 CURRENT-CARRYING CONDUCTORS IN ONE RACEWAY

Conductors must be derated according to *Notes 8* and *10, Table 310-16:*

- If more than three conductors are installed in a raceway or cable, figure 4-3(A)

- When single conductors or cable assemblies are stacked or bundled as in a cable-tray for lengths over 24 inches without spacing, figure 4-3(B)

- If the number of cable assemblies (i.e. non-metallic sheathed cable) are run together for distances over 24 inches, figure 4-3(C)

It is important to note that the Code refers to current-carrying wires for the purpose of derating when more than 3 conductors are installed in a raceway or cable. Following are the basic rules:

- DO count all current-carrying wires, figure 4-4.

- DO count neutrals of a 3-wire circuit or feeder when the system is 4-wire, 3-phase, wye connected, figure 4-5.

- DO count neutrals of a 4-wire, 3-phase, wye connected circuit or feeder when the major portion of the load is electric discharge lighting (fluorescent, mercury vapor, high pressure sodium, etc.), data processing, and other loads where third harmonic currents flow in the neutral, figure 4-6.

- DO NOT count neutrals of a 3-wire, 1-phase circuit or feeder where the neutral carries only the unbalanced current of the "hot" phase conductors, figure 4-7.

- DO NOT count equipment grounding conductors that are run in the same conduit with the circuit conductors, figure 4-8. However, the grounding conductors must be included when calculating conduit fill.

- DO NOT derate conductors in short sections of raceway 24 inches or less in length, figure 4-9.

Example 1:

Consider that 4 current-carrying No. 3 RHW copper conductors are to be installed in one raceway in an ambient temperature of 30°C (86°F). The maximum permitted load current for these conductors is determined as follows:

- The ampacity of No. 3 RHW copper conductors from *Table 310-16* is 100 amperes.

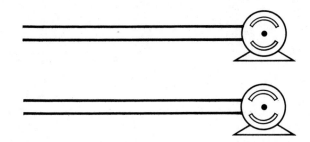

Fig. 4-4 All four of these conductors are current-carrying conductors.

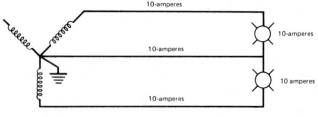

Fig. 4-5 The neutral of a 3-wire circuit supplied by a 4-wire, 3-phase, wye connected system will carry the vector sum of the "hot" conductors of the circuit. All three of these conductors are considered by the Code to be current-carrying conductors.

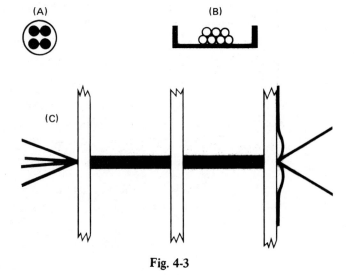

Fig. 4-3

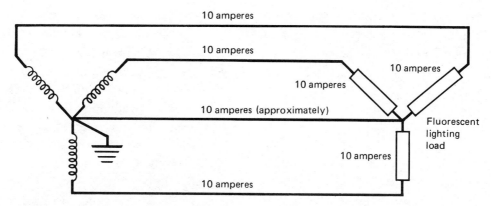

Fig. 4-6 The neutral of a 4-wire, 3-phase circuit supplying electric discharge lighting will carry approximately the same current as the "hot" conductors of the circuit. All four of these conductors are considered by the Code to be current-carrying conductors.

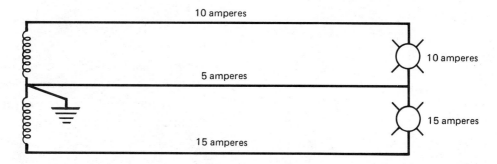

Fig. 4-7 The neutral in the diagram is not to be counted as a current-carrying conductor.

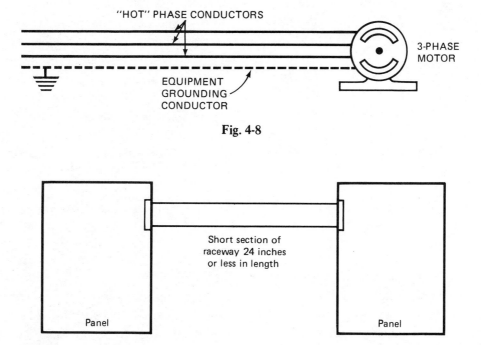

"HOT" PHASE CONDUCTORS

EQUIPMENT GROUNDING CONDUCTOR

3-PHASE MOTOR

Fig. 4-8

Short section of raceway 24 inches or less in length

Panel

Panel

Fig. 4-9 It is not necessary to derate the conductors in short sections of raceway as illustrated in the diagram, but all of the conductors must be included when calculating conduit fill.

- Apply the derating factor for 4 current-carrying conductors in one raceway:

$$100 \times 0.80 = 80 \text{ amperes}$$

Therefore, 80 amperes is the maximum permitted load current for the example given.

Example 2:

Another example is to study the branch-circuits for the drugstore where the pharmacy area lighting contains three "hot" wires and the neutral of a 4-wire, 3-phase wye connected system. The entire load is electric discharge lighting. To comply with *Note 10*, all four conductors are considered to be current-carrying conductors, thus *Note 8* derating factors must be applied. As a result, the load rated current of the four conductors in the pharmacy lighting conduit must be reduced to 80% of the ampacities as shown in *Table 310-16*. In other words, the No. 12 AWG Type THW copper conductors must be limited to:

$$25 \times 0.80 = 20 \text{ amperes}$$

This is the same as $120 \times 20 = 2,400$ volt-amperes. Since the loads are 1,200, 1,200, and 1,500 volt-amperes, they are all below the reduced allowable current limitation. The footnote to *Table 310-16* indicates that the overcurrent protection for No. 12 AWG copper wire is 20 amperes. Special considerations must be observed for No. 14, No. 12, and No. 10 AWG conductors.

An important footnote below *NEC Table 310-16* must be observed when installing No. 14, No. 12, and No. 10 AWG conductors. Notice in the table many obelisk symbols (†) (also called daggers) referring to these sizes of conductors. As can be seen, the maximum size overcurrent protection is less than the ampacities listed in the *NEC* table, as shown in table 4-2A.

Example:

To illustrate derating and maximum permitted load currents, suppose that eight (8) THW copper current-carrying conductors are to be installed in the same conduit. The connected load will be 18 amperes. The size wire to use may be determined as follows.

- No. 14 THW copper wire
  a. The maximum permitted overcurrent protection from table 4-2A (footnote to *Table 310-16*) is 15 amperes.
  b. The ampacity of No. 14 THW copper wire from *Table 310-16* is 20 amperes.
  c. Apply the derating factor (8 current-carrying wires in the raceway) from table 4-2B. (*Note 8* to *Table 310-16*).

  $$20 \times 0.70 = 14 \text{ amperes}$$
  maximum permitted load

  Thus, No. 14 THW wire is *too small* to supply the 18-ampere load.

- No. 12 THW copper wire
  a. The maximum permitted overcurrent protection from table 4-2A (footnote to *Table 310-16*) is 20 amperes.
  b. The ampacity of No. 12 THW copper wire from *Table 310-16* is 25 amperes.
  c. Apply the derating factor (8 current-carrying wires in the raceway) from table 4-2B (*Note 8* to *Table 310-16*).

  $$25 \times 0.70 = 17.5 \text{ amperes}$$
  maximum permitted load

  Thus, No. 12 THW is *too small* to supply the 18-ampere load.

- No. 10 THW copper wire
  a. The maximum permitted overcurrent protection from table 4-2A (footnote to *Table 310-16*) is 30 amperes.
  b. The ampacity of No. 10 THW copper wire from *Table 310-16* is 35 amperes.
  c. Apply the derating factor (8 current-carrying wires in the raceway) from table 4-2B (*Note 8* to *Table 310-16*).

  $$35 \times 0.70 = 24.5 \text{ amperes}$$
  maximum permitted load

  Thus, No. 10 THW copper wire is adequate to supply the 18-ampere load.

Double Derating

Double derating is not necessary. *Exception No. 2* to *Note 8* of *Table 310-16* clearly states that if a conductor has been derated because more than 3 current-carrying wires are installed in the same raceway, then it is *not* necessary to derate again because of "continuous loading," as required by *Sections 210-22(c), 220-2(a),* and *220-10(b).*

Continuous loading occurs when the circuit's maximum permitted load continues for periods of three hours or more. See definition in *NEC.* Always apply the correction factors when high temperatures are involved. The correction factors for high temperature are independent of the derating rule for more than 3 conductors in one raceway and the derating rule for continuous loads.

EXAMPLE OF HIGH TEMPERATURE AND MORE THAN 3 CONDUCTORS IN ONE RACEWAY

Consider four current-carrying No. 3 RHW copper conductors are to be installed in one raceway in an ambient temperature of 35°C (95°F). The maximum permitted load current for these conductors is determined as follows.

- The ampacity of No. 3 RHW copper conductors from *Table 310-16* = 100 amperes.

TABLE 4-2A

Wire Size	Maximum Overcurrent Protection	
	Copper	Aluminum or Copper-clad Aluminum
No. 14 AWG	15 amperes	No. 14 AWG aluminum (*not* permitted)
No. 12 AWG	20 amperes	15 amperes
No. 10 AWG	30 amperes	25 amperes

TABLE 4-2B

DERATING FACTORS (Note 8, Tables 310-16 through 310-19)	
Number of Current-carrying Conductors in Raceway	Percent of Values in *Tables 310-16* and *310-18* as Adjusted for Ambient Temperature as Necessary
4 thru 6	80 percent
7 thru 24	70 percent
25 thru 42	60 percent
43 and above	50 percent

- Apply the correction factor for 35°C (95°F).

  100 X 0.88 = 88 amperes

- Apply the derating factor for four current-carrying wires in one raceway.

  88 X 0.80 = 70.4 amperes

Therefore, 70.4 amperes is the maximum permitted load current for the example given.

DETERMINING PROPER SIZE AND TYPE CONDUCTORS

STEP 1 | What is the load current in amperes? What type of load is it? Is it a motor, non-motor, continuous, noncontinuous, transformer, electric heat, or air-conditioning? Refer to specific chapters, articles, and sections of the Code.

STEP 2 | Select conductor type (copper, aluminum, or copper-clad aluminum). Will the conductors be installed in wet or dry location, underground, conduit, or open air? Refer to *Article 310* of the Code.

STEP 3 | Refer to *Table 310-16* through *310-19* and locate conductor type and insulation type. Select tentative size. In general, conductors must have an ampacity of not less than the branch-circuit rating, and not less than the load to be served.

STEP 4 | Apply correction factors to results of Step 3, if applicable due to high ambient temperature. See footnotes to *Table 310-16* through *310-19.*

STEP 5 | Select one of the following derating factors (a, b, c, or d). No need to double derate; use only one.

a. If more than 3 current-carrying conductors in raceway or cable, derate the results of Step 4 according to *Note 8, Tables 310-16* through *310-19* (80%, 70%, 60%, 50%).

b. If load is continuous, do not exceed 80% of the branch-circuit rating, *Section 210-22(c), 220-2(a), 384-16(c).* When load is noncontinuous, the total load shall not exceed the rating of the branch-circuit.

c. For combined continuous and noncontinuous loads, conductor and overcurrent devices shall be sized for the noncontinuous load plus 125% of the continuous load.

d. For motor loads, size conductor not less than 125% of motor's full load. *Article 430* and *Section 210-22(a)*.

By completing steps 1 through 5, the maximum permitted ampacity for the conductor selected in Step 3 will be determined.

STEP 6 | Voltage drop. Calculate the voltage drop. Make sure the voltage drop does not exceed 5% (3% in branch-circuit, 2% in feeder; or 2% in branch-circuit, 3% in feeder).

STEP 7 | If Step 6 results in an acceptable conductor size that conforms to the Code, go on to the next step. If voltage drop exceeds that permitted by the Code, go back to Step 3, select a larger size conductor and then repeat Steps 4 through 9.

STEP 8 | Size overcurrent protection according to *Article 240*. Is it a motor, branch-circuit, or transformer? Refer to specific sections of the Code for the specific type of load to be served. See *Section 240-3*.

STEP 9 | Determine proper conduit size, *Tables 1 through 5, Chapter 9, National Electrical Code.*

VOLTAGE DROP

Sections 210-19(a) and *215-2(b)* of the *NEC* states that the voltage drop in a branch circuit is not to exceed 3% to the most distant outlet, figure 4-10. To determine the voltage drop, two factors must be known: (1) the resistance of the conductor, and (2) the current in the conductor. For example, the drugstore window display receptacle outlets are to be loaded to 1,500 watts or 12.5

amperes. The distance from the panelboard to the center receptacle is 85 feet (25.9 m). According to *Table 8* of the *NEC*, the resistance of a No. 12 AWG copper conductor is 1.62 ohms per 1,000 feet. Therefore, the total resistance is:

$$\frac{(1.62 \text{ ohms}/1,000 \text{ ft})}{1,000} \times 2 \times 85 \text{ ft} = 0.275 \text{ ohm}$$

The factor of 2 in the above equation is included since both the phase conductor and the neutral conductor carry current; thus, the resistance of each must be included. The voltage drop is now defined by Ohm's law.

E = IR = 12.5 amperes × 0.275 ohm = 3.44 volts

The allowable voltage drop, according to *Section 210-19(a)* is:

120 volts × 0.03 = 3.6 volts (0.03 = 3%)

This method of determining the voltage drop neglects any inductance in the circuit. However, inductance usually is insignificant at the short distances which are encountered in commercial wiring. When it is desired to calculate the voltage drop, taking into consideration both the resistance and the reactance, available tables and charts can be used to simplify the calculations. *Tables 4-3* and *4-4* show how to determine conductor sizes and voltage loss. Voltage drop is an important consideration because the higher temperature conductors now installed in electrical systems permit the use of a smaller size of conductor for a given load. These smaller conductors often result in excessive voltage loss in the circuit conductors.

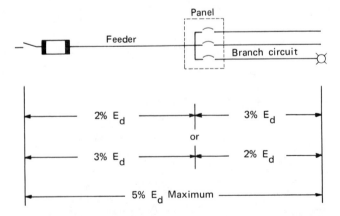

Fig. 4-10 For light, power, and lighting loads the combined voltage drop shall not exceed 5% on the feeder and branch-circuit wiring. As in the above diagram, if feeder has a 2% voltage drop, then the branch circuit shall have not over 3% voltage drop. See *Sections 210-19* and *215-2*.

RATINGS OF CONDUCTORS and TABLES to determine VOLT LOSS

With higher ratings on new insulations, it is extremely important to bear VOLT LOSS in mind, otherwise some very unsatisfactory experiences are likely to be encountered.

How to figure volt loss

MULTIPLY DISTANCE (length in feet of one wire) by the CURRENT (expressed in amperes)

by the FIGURE shown in table for the kind of current and the size of wire to be used.

THEN put a decimal point in front of the last 6 figures AND — you have the VOLT LOSS to be expected on that circuit.

Example — No. 6 copper wire in 180 feet of iron conduit — 3 phase, 40 amp. load at 80% power factor.

Multiply feet by amperes: 180 × 40 = 7200

Multiply this number by number from table for No. 6 wire three-phase at 80% power factor: 7200 × 735 = 5292000

Place decimal point 6 places to left.

This gives volt loss to be expected: *5.292 volts*

(For a 240 volt circuit the % voltage drop is $\frac{5.292}{240} \times 100$ or 2.21%.)

These tables take into consideration REACTANCE ON AC CIRCUITS as well as resistance of the wire.

Remember on short runs to check to see that the size and type of wire indicated has sufficient ampere capacity.

How to select size of wire

MULTIPLY DISTANCE (length in feet of one wire) by the CURRENT (expressed in amperes)

DIVIDE that figure INTO the permissible VOLT LOSS multiplied by 1,000,000

Look under the column applying to the type of current and power factor for the figure nearest, but not above your result

AND — you have the size of wire needed.

Example — Copper wire in 180 feet of iron conduit — 3 phase, 40 amp. load at 80% power factor — volt loss from local code equals 5.5 volts.

Multiply feet by amperes: 180 × 40 = 7200

Divide permissible volt loss multiplied by 1,000,000 by this number: $\frac{5.5 \times 1,000,000}{7200} = 764$

Select number from table, three-phase at 80% power factor that is nearest but not greater than 764. This number is 735 which indicates the size of wire needed: *No. 6*

TABLE 4-3 COPPER CONDUCTORS

| | | AMPACITY* | | | | VOLT LOSS (See Explanation Above) | | | | | | | | | |
| | | Type T, TW (60°C Wire) | Type RH THWN, RHW, THW (75°C Wire) | Type RHH, THHN, XHHW (90°C Wire) | Direct Current | THREE-PHASE — 60 Cycle, lagging power factor | | | | | SINGLE-PHASE — 60 Cycle, lagging power factor | | | | |
	WIRE SIZE					100%	90%	80%	70%	60%	100%	90%	80%	70%	60%
COPPER CONDUCTORS IN IRON CONDUIT	14	20	20	25	6100	5280	4800	4300	3780	3260	6100	5551	4964	4370	3772
	12	25	25	30	3828	3320	3030	2720	2400	2080	3828	3502	3138	2773	2404
	10	30	35	40	2404	2080	1921	1733	1540	1340	2404	2221	2003	1779	1547
	8	40	50	55	1520	1316	1224	1120	1000	880	1520	1426	1295	1159	1107
	6	55	65	75	970	840	802	735	665	590	970	926	850	769	682
	4	70	85	95	614	531	530	487	445	400	614	613	562	514	462
	3	85	100	110	484	420	425	398	368	334	484	491	460	425	385
	2	95	115	130	382	331	339	322	300	274	382	392	372	346	317
	1	110	130	150	306	265	280	270	254	236	306	323	312	294	273
	0	125	150	170	241	208	229	224	214	202	241	265	259	247	233
	00	145	175	195	195	192	166	190	188	181	173	192	219	217	199
	000	165	200	225	152	132	157	158	155	150	152	181	183	179	173
	0000	195	230	260	121	105	131	135	134	132	121	151	156	155	152
	250M	215	255	290	102	89	118	123	125	123	103	136	142	144	142
	300M	240	285	320	85	74	104	111	112	113	86	120	128	130	131
	350M	260	310	350	73	63	94	101	105	106	73	108	117	121	122
	400M	280	335	380	64	55	87	95	98	100	64	100	110	113	116
	500M	320	380	430	51	45	76	85	90	92	52	88	98	104	106
	600M	355	420	475	43	38	69	79	85	87	44	80	91	98	101
	700M	385	460	520	36	33	64	74	80	84	38	74	86	92	97
	750M	400	475	535	34	31	62	72	79	82	36	72	83	91	95
	800M	410	490	535	32	29	61	71	76	81	33	70	82	88	93
	900M	435	520	585	28	26	57	68	74	78	30	66	78	85	90
	1000M	455	545	615	26	23	55	66	72	76	27	63	76	83	88
COPPER CONDUCTORS IN NON-MAGNETIC CONDUIT Lead covered cables or installation in fibre or other non-magnetic conduit, etc.	14	15	15	15	6100	5280	4790	4280	3760	3240	6100	5530	4936	4336	3734
	12	20	20	20	3828	3320	3020	2700	2380	2055	3828	3483	3112	2742	2369
	10	30	30	30	2404	2080	1910	1713	1513	1311	2404	2202	1978	1748	1512
	8	40	45	55	1520	1316	1220	1100	976	851	1520	1406	1268	1128	982
	6	55	65	75	970	840	787	715	641	562	970	908	825	740	648
	4	70	85	95	614	531	517	466	422	374	614	596	538	486	431
	3	85	100	110	484	420	410	379	344	308	484	474	438	397	355
	2	95	115	130	382	331	326	303	278	250	382	376	350	321	288
	1	110	130	150	306	265	266	251	232	211	306	307	289	267	243
	0	125	150	170	241	208	216	206	192	176	241	249	237	221	203
	00	145	175	195	192	166	176	170	160	148	192	203	196	184	171
	000	165	200	225	152	132	145	141	134	126	152	167	163	155	145
	0000	195	230	260	121	105	119	118	114	108	121	137	136	131	125
	250M	215	255	290	102	89	105	106	104	100	103	121	122	120	115
	300M	240	285	320	85	74	92	95	93	91	86	106	109	107	105
	350M	260	310	350	73	63	82	85	84	83	73	94	98	97	96
	400M	280	335	380	64	55	75	78	79	78	64	86	90	91	90
	500M	320	380	430	51	45	64	69	71	70	52	74	80	82	81
	600M	355	420	475	43	38	57	63	66	66	44	66	73	76	76
	700M	385	460	520	36	33	53	58	61	63	38	61	67	70	73
	750M	400	475	535	34	31	51	56	60	61	36	59	65	69	70
	800M	410	490	535	32	29	49	55	58	60	33	57	64	67	69
	900M	435	520	585	28	26	46	52	55	57	30	53	60	64	66
	1000M	455	545	615	26	23	43	50	54	56	27	50	58	62	64

*These are ampacities of the conductors to which further derating and correction factors must be applied where applicable. See footnote to *NEC Table 310-16*.

Maximum overcurrent protection for copper conductors is 15 amperes for No. 14 AWG, 20 amperes for No. 12 AWG, and 30 amperes for No. 10 AWG.

TABLE 4-4 ALUMINUM CONDUCTORS

| | AMPACITY* | | | | VOLT LOSS (see Explanation Above) | | | | | | | | | | |
| | Type T, TW (60°C Wire) | Type RH, THWN, RHW, THW (75°C Wire) | Type RHH, THHN, XHHW (90°C Wire) | Direct Current | THREE PHASE — 60 Cycle, lagging power factor | | | | | SINGLE PHASE — 60 Cycle, lagging power factor | | | | |
WIRE SIZE					100%	90%	80%	70%	60%	100%	90%	80%	70%	60%
ALUMINUM CONDUCTORS IN IRON CONDUIT														
12	15	20	25	6040	5230	4760	4260	3740	3243	6040	5500	4920	4320	3745
10	25	30	35	3800	3291	3005	2690	2380	2080	3800	3470	3110	2750	2395
8	30	40	45	2390	2070	1905	1725	1525	1330	2390	2200	1990	1760	1540
6	40	50	60	1530	1325	1238	1126	1005	890	1530	1430	1300	1160	1030
4	55	65	75	966	837	795	726	647	585	966	918	838	747	675
2	75	90	100	606	526	511	473	434	397	606	590	546	498	456
1	85	100	115	480	415	414	386	355	330	480	478	446	410	380
0	100	120	135	382	331	336	317	294	277	382	388	366	340	320
00	115	135	150	302	262	272	260	244	232	302	314	300	282	268
000	130	155	175	240	210	225	217	206	199	242	260	250	238	230
0000	155	180	205	192	168	185	182	175	173	194	214	210	202	200
250M	170	205	230	161	142	163	163	157	153	164	188	188	181	177
300M	190	230	255	134	119	141	142	140	141	137	163	164	162	163
350M	210	250	280	115	102	126	128	127	125	118	146	148	147	148
400M	225	270	305	101	91	115	120	119	122	105	133	138	137	141
500M	260	310	350	80	74	100	104	106	107	85	115	120	122	124
600M	285	340	385	67	62	88	95	98	101	72	102	110	113	117
700M	310	375	420	58	55	82	88	92	97	64	95	102	106	112
750M	320	385	435	54	52	79	85	89	94	60	91	98	103	108
800M	330	395	450	50	49	76	83	87	93	57	88	96	101	107
900M	335	425	480	45	45	72	80	83	88	52	83	92	96	102
1000M	375	445	500	40	42	68	76	81	85	48	79	88	93	98
ALUMINUM CONDUCTORS IN NON-MAGNETIC CONDUIT Lead covered cables or installation in fibre or other non-magnetic conduit, etc.														
12	15	15	15	6040	5230	4750	4250	3720	3217	6040	5490	4900	4300	3715
10	25	25	25	3800	3290	3000	2680	2360	2040	3800	3460	3100	2730	2360
8	30	40	45	2390	2070	1900	1701	1501	1304	2390	2190	1970	1740	1510
6	40	50	60	1530	1325	1230	1110	990	866	1530	1420	1280	1140	1000
4	55	65	75	966	837	787	715	641	570	966	908	826	740	656
2	75	90	100	606	525	504	462	419	378	606	580	534	484	435
1	85	100	115	480	416	405	376	343	312	480	468	434	396	360
0	100	120	135	382	331	328	307	282	258	382	378	354	326	299
00	115	135	150	302	262	265	251	232	216	302	306	290	268	249
000	130	155	175	240	208	217	206	175	164	240	250	238	202	189
0000	155	180	205	192	166	177	171	161	154	192	204	197	186	176
250M	170	205	230	161	139	153	151	144	138	161	177	174	166	159
300M	190	230	255	134	116	133	132	127	125	134	153	152	147	144
350M	210	250	280	115	100	117	117	114	114	115	135	135	132	131
400M	225	270	305	101	87	106	107	106	105	101	122	124	122	121
500M	260	310	350	80	70	89	92	92	91	81	103	106	106	105
600M	285	340	385	67	59	79	83	83	83	68	91	96	96	96
700M	310	375	420	58	50	71	76	78	82	58	82	88	90	94
750M	320	385	435	54	48	68	73	75	76	55	79	84	87	88
800M	330	395	450	50	44	66	71	73	74	51	76	82	84	86
900M	355	425	480	45	40	61	67	69	71	46	70	77	82	82
1000M	375	455	500	40	36	57	63	66	67	41	66	73	76	78

*These are ampacities of the conductors, to which further derating and correction factors must be applied where applicable. See footnote to *NEC Table 310-16.*

Maximum overcurrent protection for aluminum or copper-clad aluminum conductors is 15 amperes for No. 12 AWG, and 25 amperes for No. 10 AWG.

ENERGY SAVINGS CONSIDERATIONS

The rapidly escalating cost of energy encourages the consideration to install on size larger conductors than is required by the Code. This is particularly true in the smaller size conductors, i.e. #14, #12, #10, and #8. Energy savings are minimal in the larger conductor sizes 1/0 and larger.

Heat (watts, energy) equals amperes times amperes times the resistance. The watts loss (heat) in the copper conductors can be calculated as follows:

$$\text{Watts} = I^2 R$$

Looking at a typical circuit, energy cost comparisons can be made.

PANEL — 100 feet / #12 AWG — Load 16 Amperes

Resistance value of #12 copper, from *Table 8, NEC,* equals 0.162 ohm per 100 feet. For #1 copper, the resistance value is 0.1018 ohm per 100 feet.

Example 1:

Using the #12 copper wire, the total watts loss in the conductor is:

$$\text{Watts} = I^2 R$$
$$= 16 \times 16 \times 0.162 \times 2 \text{ wires}$$
$$= 82.9 \text{ Watts}$$

$$\text{kWh per year} = \frac{82.9 \times 10 \text{ hours per day} \times 250 \text{ days per year}}{1,000}$$
$$= 207.25 \text{ kWh per year}$$

Cost of copper loss @ 8¢ per kWh equals $16.58 per year

Example 2:

Using #10 copper wire, the total watts loss in the conductor is:

$$\text{Watts} = I^2 R$$
$$= 16 \times 16 \times 0.1018 \times 2 \text{ wires}$$
$$= 52.1 \text{ Watts}$$

$$\text{kWh per year} = \frac{52.1 \times 10 \text{ hours per day} \times 250 \text{ days per year}}{1,000}$$

$$= 130.25 \text{ kWh per year}$$

Cost of copper loss @ 8¢ per kWh equals $10.42 per year

Comparison:

If list price of No. 12 TW per 1,000 feet is $66.46, the cost of 200 feet is $13.29.

If list price of No. 10 TW per 1,000 feet is $106.40, the cost of 200 feet is $21.28. Thus,

No. 12 TW Circuit
Cost of wire	$13.29
Annual watts loss cost	$16.58
	$29.87

No. 10 TW Circuit
Cost of wire	$21.28
Annual watts loss cost	$10.42
	$31.70

The first year shows that the initial cost of both circuits are almost the same. Note that the added cost of the larger wire is $7.99; the copper watts loss savings is $6.16. The first year of operation is a "break even" year. Thereafter, the energy savings is $6.16 on an annual basis. That is a significant return on an initial investment of just $7.99.

If the electrical contractor installs 10,000 feet of No. 10 TW instead of 10,000 feet of No. 12 TW, a considerable energy savings would be experienced. However, the contractor must consider whether or not the Code requires the use of larger size conduit. (See *Chapter 9* of the *NEC*.)

The larger size conductors also keep voltage drop to a minimum which permits the connected electrical equipment to operate more efficiently.

Connections for the wires on the wiring devices terminal could be more time consuming. The electrician might prefer to attach #12 AWG pigtails to the terminals of wiring devices and then splice these to the #10 circuit conductors.

Consideration must also be given to the reduction of heat contributed by the larger conductors to the air-conditioning system, or its heat contribution to the heating system during the heating season.

The savings in energy consumption through the use of larger conductors is worthy of consideration.

RECEPTACLES

The National Electrical Manufacturers Association (NEMA) has developed standards for the physical appearance of locking and nonlocking plugs and receptacles. The differences in the plugs and receptacles are based on the ampacity and voltage rating of the device. For example, the two most commonly used receptacles are the NEMA 5-15R, figure 4-11, and the NEMA 5-20R, figure 4-12. The NEMA 5-15R receptacle has a 15-ampere, 125-volt rating and has two parallel slots and a ground pinhole. This receptacle will accept the NEMA 5-15P plug only, figure 4-13. The NEMA 5-20R receptacle has two parallel slots and a "T" slot. This receptacle is rated at 20 amperes, 125 volts. The NEMA 5-20R will accept either a

Fig. 4-11 NEMA 5-15R.

NEMA 5-15P or 5-20P plug, figure 4-14. A NEMA 6-20R receptacle is shown in figure 4-15. This receptacle has a rating of 20 amperes at 250 volts and will accept either the NEMA 6-15P or 6-20P plug, figure 4-16. The acceptance by most of the receptacles of 15- and 20-ampere plugs complies with *Table 210-21(b)(3)* of the *NEC* which permits the installation of either the NEMA 5-15R or 5-20R receptacle on a 20-ampere branch circuit. Another receptacle which is specified for the commercial building is the NEMA 6-30R, figure 4-17. This receptacle is rated at 30 amperes, 250 volts.

The NEMA standards for general-purpose nonlocking plugs and receptacles are shown in figure 4-18. A special note should be made of the differences between the 125/250 (NEMA 14) devices and the 3φ, 250-volt (NEMA 15) devices. The connection of the 125/250-volt receptacle requires a neutral, a grounding wire, and two-phase connections. For the 3φ, 250-volt receptacle, a grounding wire and three-phase connections are required.

GROUND-FAULT CIRCUIT INTERRUPTERS *(NEC Section 210-8)*

Section 210-8 of the Code discusses ground-fault circuit interrupters (GFCI) when installed in dwellings, hotels and motels. The Code does not require GFCI protection for certain branch circuits in commercial and industrial facilities. Future editions of the Code may require GFCI protection for certain branch circuits in commercial and/or industrial facilities.

Electrocutions and personal injury have resulted because of electrical shock from appliances

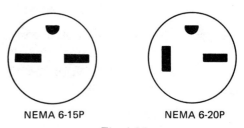

Fig. 4-15 NEMA 6-20R.

NEMA 6-15P   NEMA 6-20P

Fig. 4-16

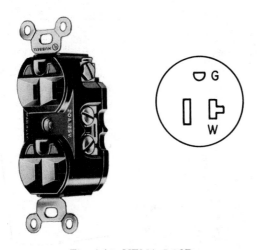

Fig. 4-12 NEMA 5-20R.

Fig. 4-13 NEMA 5-15P.

Fig. 4-14 NEMA 5-20P.

Fig. 4-17 NEMA 6-30R.

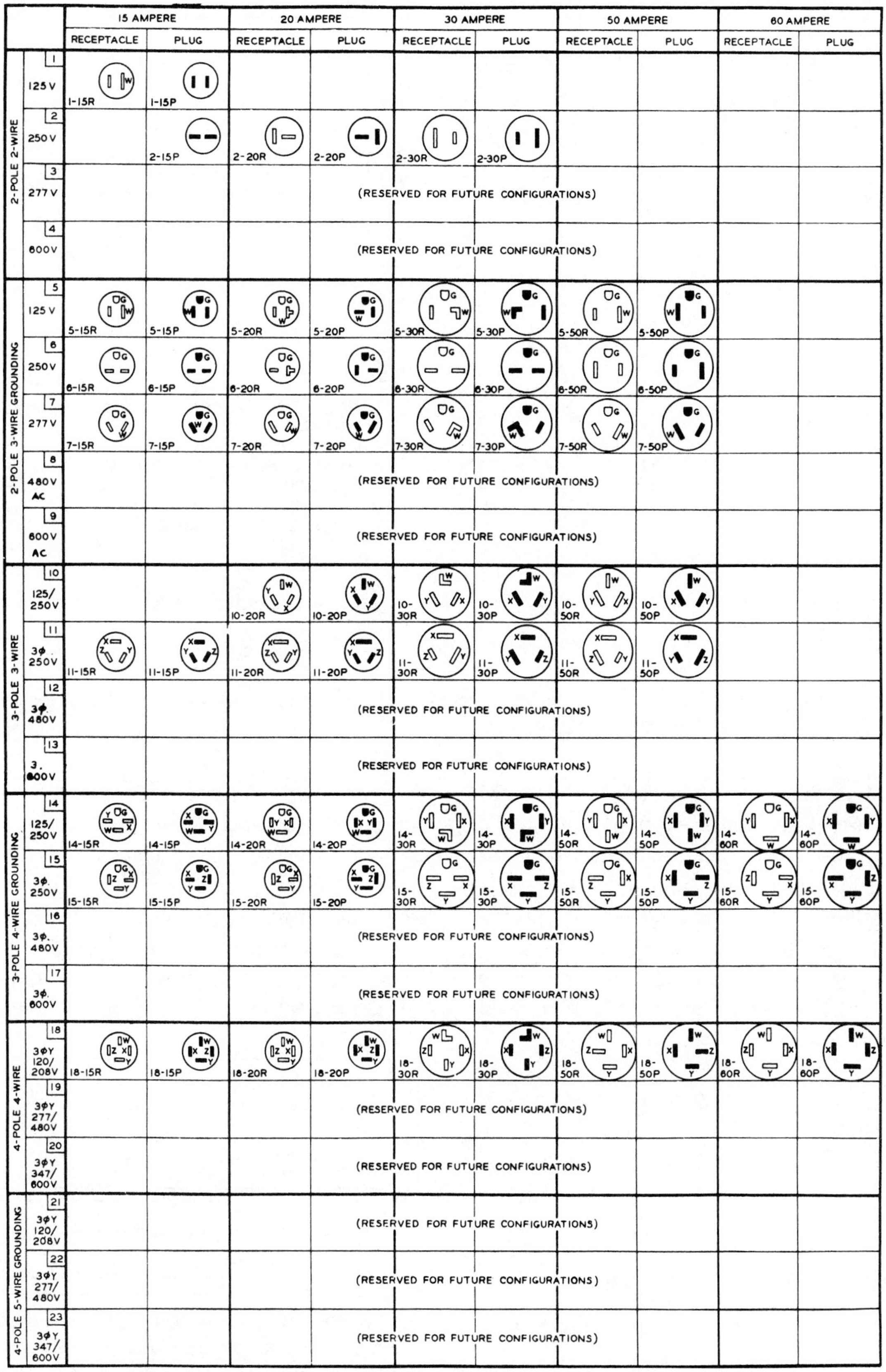

Fig. 4-18 NEMA configurations for general-purpose nonlocking plugs and receptacles.

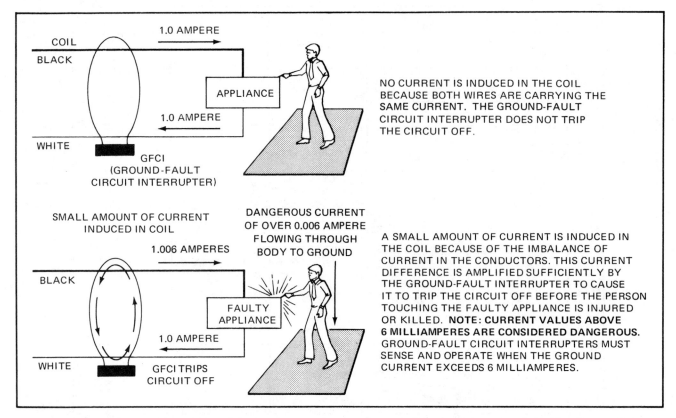

Fig. 4-19 Basic principle of how a ground-fault circuit interrupter operates. Reprinted from Ray C. Mullin, *Electrical Wiring – Residential,* figure 6-8. © 1984 by Delmar Publishers Inc.

such as radios, shavers, and electric heaters. This shock hazard exists whenever a person touches both the defective appliance and a conducting surface such as a water pipe, metal sink or any conducting material that is grounded. To protect against this possibility of shock, 15- and 20-ampere branch circuits can be protected with GFCI receptacles or GFCI circuit breakers. The receptacles are the most commonly used.

The Underwriters Laboratories require that Class A GFCIs trip on ground-fault currents of 4 to 6 milliamperes (0.004 to 0.006 ampere). Figure 4-19 illustrates the principle of how a GFCI operates.

ELECTRIC BASEBOARD HEATERS

Some buildings are heated with electric baseboard heaters. It is not the intent of this book to provide the student with the whys and wherefores of electric heat, but to provide the information necessary to meet the Code regarding the location of receptacle outlets when electric baseboard heaters are installed.

Electric baseboard heaters are available with or without receptacle outlets. Figure 4-20 shows the relationship of an electric baseboard heater to a receptacle outlet. *Article 424* of the Code states the Code requirements for fixed electric space heating equipment.

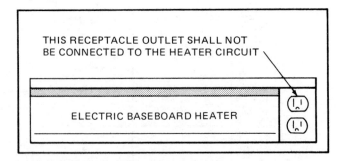

Fig. 4-20 Factory-mounted receptacle outlets on permanently installed electric baseboard heaters may be counted as the required outlets for the space occupied by the baseboard unit. See *NEC Section 210-52(a).* Reprinted from Ray C. Mullin, *Electrical Wiring – Residential,* figure 23-6. © 1984 by Delmar Publishers Inc.

——REVIEW———————————————

Note: Refer to the *National Electrical Code* or the plans as necessary.

1. For a typical beauty parlor with an area of 700 square feet, (65.1 square meters), a general lighting load of not less than _____ watts shall be included for branch circuits.

2. An air-conditioning system in a typical beauty salon has the following rated load currents:

    Compressor-motor   14.1 amperes, 208 volts, three phase
    Condenser-evaporator (one motor)   3.3 amperes, 208 volts, single phase

    a. The branch-circuit loading allowance is _____ volt-amperes.

    b. The actual current in each phase will be _____ , _____ , _____ and _____ ampere(s).

    c. Neglecting voltage drop, the minimum permitted branch-circuit conductor size is No. _____ AWG Type TW.

    d. The branch-circuit short-circuit protection should be _____ amperes maximum.

3. Three 7,280-watt, 208-volt, single-phase loads are located 100 feet (30.5 m) from the panelboard. Three separate branch circuits are to be installed in conduit to serve these loads.

    a. The current in each circuit is _____ amperes.

    b. The conductors shall be not less than _____ AWG, type _____ .

    c. The overcurrent device rating shall be _____ amperes.

    d. The voltage drop will be _____ %.

    Using the information in figure 4-18, select by number the correct receptacle for each of the following outlets.

4. 120/208-volt, single-phase, 20-ampere outlet _____

5. 230-volt, three-phase, 50-ampere outlet _____

6. 208-volt, single-phase, 30-ampere outlet _____

7. When Type THHN conductors are installed in an ambient temperature of 90°F, the ampacity as shown in *Table 310-16* must be reduced to _____ .

8. When four Type THW, current-carrying conductors are installed in one raceway, the ampacity as shown in *Table 310-16* must be reduced to _____ .

9. When four No. 1, Type THHN copper current-carrying wires are installed in a location where the ambient temperature will be 110°F continuously, the true ampacity of the conductors is: (underline one)

    a. 120 amperes   b. 104 amperes   c. 130 amperes

—UNIT 5—

Low-voltage Remote-control Lighting

— OBJECTIVES —

After completing the study of this unit, the student will be able to

- list the components of a low-voltage remote-control wiring system.
- select the appropriate *NEC Sections* governing the installation of a low-voltage remote-control wiring system.
- demonstrate the correct connections for wiring a low-voltage remote-control system.

LOW-VOLTAGE REMOTE-CONTROL LIGHTING

Conventional general lighting control is used in the major portion of the commercial building. For the drugstore, however, a method known as *low-voltage remote-control lighting* is selected because of the number of switches required and because it is desired to have extensive control of the lighting.

Relays

A low-voltage remote-control wiring system is relay operated. The relay is controlled by a low-

voltage switching system and in turn controls the power circuit connected to it, figure 5-1. The low-voltage, split coil relay is the heart of the low-voltage remote-control system, figure 5-2. When the *On* coil of the relay is energized, the solenoid mechanism causes the contacts to move into the *On* position to complete the power circuit. The contacts stay in this position until the *Off* coil is energized. When this occurs, the contacts are withdrawn and the power circuit is opened. The red wire is the *On* wire; the black wire is *Off*, and the blue wire is common to the transformer.

The low-voltage relay is available in two mounting styles. One style of relay is designed to

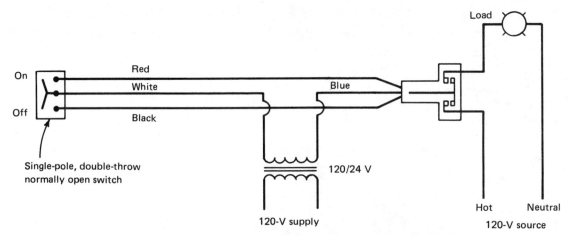

Fig. 5-1 Basic connection diagram for low-voltage remote-control system.

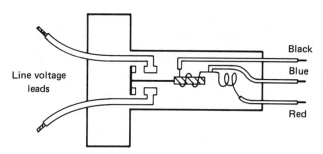

Fig. 5-2 Low-voltage relay.

Fig. 5-3 1/2-inch relay and 3/4-inch rubber grommet.

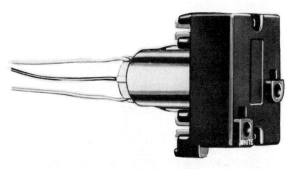

Fig. 5-4 Plug-in relay.

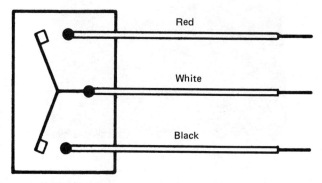

Fig. 5-5 Single-pole, double-throw, normally open, low-voltage control switch.

mount through a 1/2-inch knockout opening, figure 5-3. For a 3/4-inch knockout, a rubber grommet is inserted to isolate the relay from the metal. This practice should insure quieter relay operation. The second relay mounting style is the *plug-in relay*, figure 5-4. This type of relay is used in an installation where several relays are mounted in one enclosure. The advantage of the plug-in relay is that it plugs directly into a bus bar. As a result, it is not necessary to splice the line voltage leads. However, since the equipment is constructed so that each branch circuit is to supply three relays, some space in the equipment will be unusable in situations where the branch circuit supplies only one relay, such as circuits 9, 11, and 13 in the drugstore.

Single Switch

The switch used in the low-voltage remote-control system is a normally open, single-pole, double-throw, momentary contact switch, figure 5-5. This switch is approximately one-third the size of a standard single-pole switch. In general, this type of switch has short lead wires for easy connections, figure 5-6.

To make connections to the low-voltage switches, the white wire is common and is connected to the 24-volt transformer source. The red wire connects to the *On* circuit and the black wire connects to the *Off* circuit.

Master Control

It is common practice to group low-voltage switches to provide control of several circuits from

Fig. 5-6 Low-voltage switch.

a single point. Figure 5-7 shows a simple grouping in which several switches are ganged together. A master switching control has a rotary switch device with 12 positions, figure 5-8. The dial of the master control is depressed to make contact. While depressed, the dial can be rotated through all of its positions to achieve a rapid control of the connected lighting circuits.

A more elaborate system of switching is possible through the use of a *motorized master control*, figure 5-9. This device permits the control of up to 25 circuits at one time. For each touch of the *On* or *Off* position of a standard low-voltage switch, the master control makes a complete sweep of the 25 positions to cause the operation of the relays connected to these positions. Normally, one motorized master unit controls the *On* positions and a second motorized unit controls the *Off* positions.

Master Control with Rectifiers

Several relays can be operated individually or from a single switch through the use of *rectifiers*. The principle of operation of a master control with rectifiers is based on the fact that a rectifier permits current in only one direction, figure 5-10.

For example, if two rectifiers are connected as shown in figure 5-11, current cannot exist from A to B, or from B to A; however, current can exist

Fig. 5-7 Eight-switch master control.

Fig. 5-8 Rotary master switch.

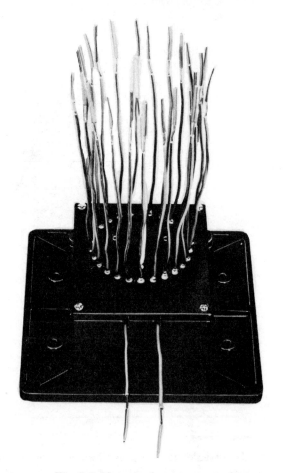

Fig. 5-9 Motorized master control.

from A to C, or from B to C. Thus, if a rectifier is placed in one lead of the low-voltage side of the transformer, and additional rectifiers are used to isolate the switches, then a master switching arrangement is achieved, figure 5-12. This method of master control is used in the drugstore. Although

a switching schedule is included in the specifications, the electrician may find it necessary to prepare a connection diagram similar to that shown in figure 5-13. The relays and rectifiers, figure 5-14, for the drugstore master control are located in the low-voltage control panel.

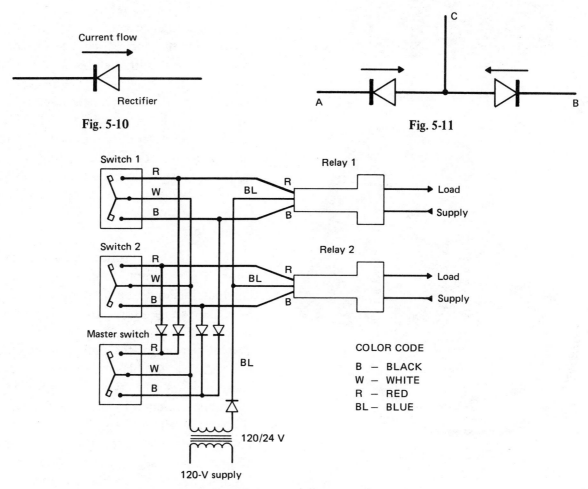

Fig. 5-10

Fig. 5-11

Fig. 5-12 Master control with rectifiers.

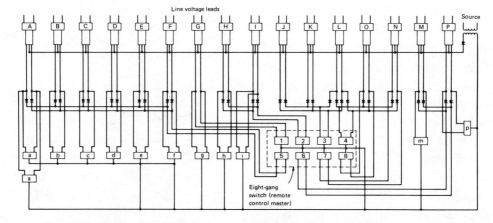

Fig. 5-13 Connection diagram for drugstore low-voltage remote-control wiring.

Fig. 5-14 Rectifier.

Fig. 5-15 35-VA transformer.

WIRING METHODS

A separate article of the *NEC* (*Article 725, Remote-Control and Signal Circuits*) governs the installation of a low-voltage system. *Article 725* provisions apply to remote-control circuits, low-voltage relay switching, low-energy power circuits, and low-voltage circuits.

The drugstore low-voltage wiring is classified as a Class 2 circuit, *Table 725-31(a)*. Since the power source of the circuit is limited (by the definition of a low-voltage circuit), overcurrent protection is not required.

The circuit transformer, figure 5-15, is designed so that in the event of an overload, the out-put voltage decreases and there is less current output. Any overload can be counteracted by these energy-limiting characteristics through the use of a specially designed transformer core. If the transformer is not self-protected, *Table 725-31(a)*, a thermal device may be used to open the primary side to protect the transformer from overheating. This thermal device resets automatically as soon as the transformer cools. Other transformers are protected with nonresetting fuse links or externally mounted fuses. Although the *NEC* does not require that the low-voltage wiring be installed in a raceway, the specifications for the commercial building do contain this requirement. The advantage of using a raceway for the installation is that it provides a means for making future additions at a minimum cost. A disadvantage of this approach is the initial higher construction cost.

Conductors

No. 18 AWG solid conductors are used for low-voltage, remote-control systems. Larger conductors should be used for long runs to minimize the voltage drop. The cables for the installation usually contain two or three conductors. The insulation on the individual conductors may be either a double-cotton covering or plastic. Regardless of the type of insulation used the individual conductors in the cables can be identified readily, figure 5-16. To simplify connections, low-voltage cables are available in various color combinations, such as blue-white, red-black, or black-white-red. To install these color-coded cables correctly throughout an entire installation, the wires are connected like color to like color. A cable suitable for outdoor use, either overhead or underground, is available for low-voltage, remote-control systems.

Low-voltage Panel

The low-voltage relays in the drugstore installation are to be mounted in an enclosure next to the power panel, figure 5-17. A barrier in this low-voltage panel separates the 120-volt power lines from the low-voltage control circuits, in compliance with *Section 725-38*.

Fig. 5-16 Low-voltage control wire.

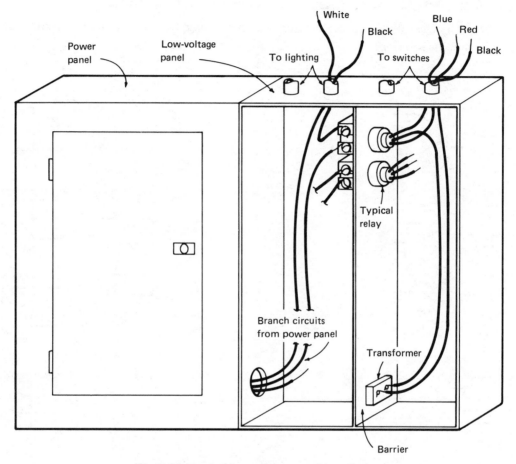

Fig. 5-17 Typical low-voltage panel installation.

REVIEW

Note: Refer to the *National Electrical Code* or the plans as necessary.

1. The low-voltage relay has a _____ coil design.

2. If the relay is in the *On* position and the *On* circuit is re-energized, the relay (will) (will not) change to the *Off* position. (Underline the correct answer.)

3. Match the following items by placing the letter of the correct wire in the appropriate blank of the column on the left.

    a. _____ Relay *On* circuit   A. Red wire
    b. _____ Relay common circuit   B. Black wire
    c. _____ Relay *Off* circuit   C. Blue wire
    d. _____ Switch *On* circuit   D. White wire
    e. _____ Switch *Off* circuit   E. Green wire
    f. _____ Switch common circuit   F. Yellow wire

4. Indicate the current direction for the device shown by the following symbol: ——▷|——

_____

5. What *Article* of the *NEC* governs the installation of remote-control and signal circuits?

_____

6. Low-voltage wiring (must) (need not) be installed in raceway to comply with the *NEC*. (Underline the correct answer.)

7. Complete the following connection diagram *using rectifiers* for the master control. Indicate wire colors.

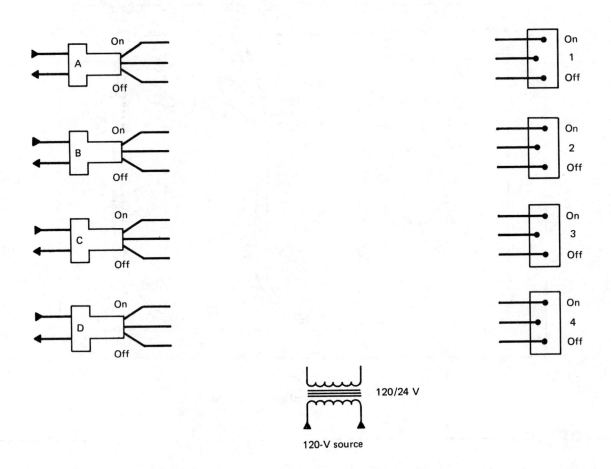

Switching schedule	
Switch	Control relay
1	A
2	B
3	C & D
4	A & B

UNIT 6

Reading Electrical Drawings— Bakery

OBJECTIVES

After completing the study of this unit, the student will be able to

- determine loading requirements.
- select correct conduit sizes.
- select the correct branch-circuit conductors.
- lay out the conduit installation.

PRINTS

Rarely is a perfect set of electrical drawings prepared for use by the electrician. Regardless of the care taken with the drawings, a number of errors usually can be expected because of the complexity of the work. Therefore, it is recommended that the electrician check the drawings to uncover any serious discrepancies. This procedure also helps the electrician to become familiar with the plans. Since the electrician may be required to make such decisions as wire and conduit sizing, these decisions are best made before the work starts. This unit provides experience in reviewing plans and determining the job requirements.

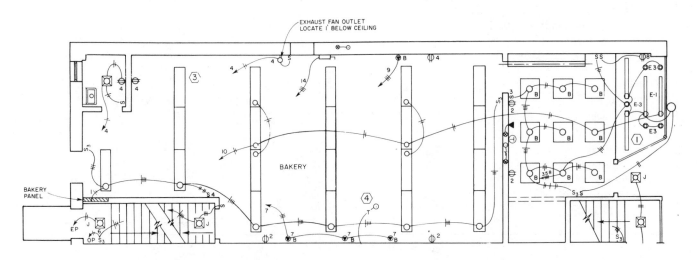

NOTE: FOR COMPLETE BLUEPRINT, REFER TO BLUEPRINT E-2.

APPLICATION QUESTIONS

1. Complete the following loading schedule for the bakery using the drugstore loading schedule in Unit 4, page 46, as a guide. Refer to the *National Electrical Code* or the plans as necessary.

BAKERY LOADING SCHEDULE				
Loading	NEC Article	NEC Minimum	Actual Allowance	Design Value
GENERAL LIGHTING:				
Store Area:				
1,200 sq. ft. X _____ watts/sq. ft.	220-2(b)	_____		
_____ Style C luminaires @ 100 volt-amperes			_____	
_____ Style B luminaires @ 120 volt-amperes			_____	
_____ Style J luminaires @ 150 watts			_____	
Storage Area:				
1,200 sq. ft. X _____ watts/sq. ft.	220-2(b)	_____		
_____ Style L luminaires @ 100 volt-amperes			_____	
Total general lighting				
Value to be used				_____
SHOW WINDOW:				
13.5 ft. X _____ watts/ft.	220-2(c)	_____		
_____ Style E1 fixtures @ 60 volt-amperes	Ex. No. 3		_____	
_____ Style E3 fixtures @ 150 watts			_____	
_____ receptacle outlets @ 1,500 watts			_____	
Total show window				
Value to be used				_____
MOTOR LOAD:				
Exhaust fan _____ volts X 2.9 amperes		348	348	
Cake mixer 208 volts X 1.73 X 3.96 amperes		1,426	1,426	
Cake mixer 208 volts X 1.73 X 7.48 amperes		2,693	2,693	
Dough divider 208 volts X 1.73 X 2.2 amperes		792	792	
Doughnut machine 208 volts X 1.73 X 2.2 amperes		792	792	
plus heater		2,000	2,000	
25% of largest motor load				
(2,693 volt-amperes X 0.25 _____)	430-24	673		
Total motor load		8,724	8,051	
Value to be used				8,724
OTHER:	220-2(c)			
Bake oven		16,000	16,000	
Receptacle outlets:				
_____ outlets X _____		_____		
Sign outlet:				
1 outlet at 100 volt-amperes		100	_____	
Total other load				
Total loading allowance				
Value to be used				_____

Questions 2 through 9 are concerned with branch circuits 3, 5, and 8 (see Panel Schedule in specifications). Assume that the distances for the voltage-drop calculations are 80 ft. (24.4 m), 90 ft. (27.4 m), and 100 ft. (30.5 m) respectively for the branch circuits.

2. The rated load current in circuit No. 8 is _____ amperes.

3. The voltage drop in circuit No. 8 using No. 12 copper wire is _____ volts.

4. The permissible voltage drop for circuit No. 8 is _____ volts.

5. The correct conductor size for circuit No. 8 is No. _____ .

6. The correct conductor size for circuit No. 3 is No. _____ .

7. The correct conductor size for circuit No. 5 is No. _____ .

8. Using type RHW conductors, the total cross-sectional area is _____ square inches, requiring _____-inch conduit.

9. Using type THHN conductor, the total cross-sectional area is _____ square inches, requiring _____-inch conduit.

10. On the following diagram of the layout of the bakery, indicate the conduit routing and size, and the conductor number and size. When making the decision on how to lay out the conduit installation, the electrician should consider (1) the amount of material required, (2) the ease of installation, and (3) the electrical characteristics of the circuit.

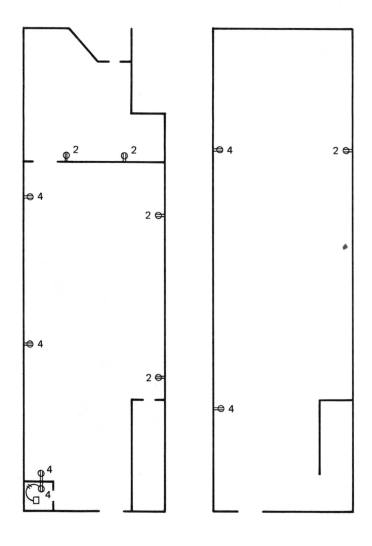

═══UNIT 7═══

Switch Control of Lighting Circuits

═══ OBJECTIVES ═══

After completing the study of this unit, the student will be able to

- select the switch with the proper rating for a particular application.
- connect single-pole, three-way, four-way, and double-pole switches into control circuits.

During the course of the work, an electrician installs and connects numerous switches of various types. Therefore, it is essential for the electrician to know how each type of switch operates, and what are the standard connections for each type. In addition, the electrician must understand the significance of the current and voltage ratings of the switches used in typical installations.

SWITCHES

Underwriters Laboratories classifies toggle switches used for lighting circuits as *general-use snap switches*, figure 7-1. These switches are divided into two categories.

Category 1 contains those ac/dc general-use snap switches which are used to control:

- alternating-current or direct-current circuits.
- resistive loads not to exceed the ampere rating of the switch at rated voltage.
- inductive loads not to exceed one-half the ampere rating of the switch at rated voltage.
- tungsten filament lamp loads not to exceed the ampere rating of the switch at 125 volts when marked with the letter "T." (A tungsten filament lamp draws a very high current at the instant the circuit is closed. As a result, the switch is subjected to a severe current surge.)

The ac/dc general-use snap switch normally is not marked *ac/dc*. However, it is always marked with the current and voltage rating, such as *10A-125V,* or *5A-250V-T.*

Category 2 contains those ac general-use snap switches which are used to control:

- alternating-current circuits only.
- resistive, inductive, and tungsten-filament lamp loads not to exceed the ampere rating of the switch at 120 volts.

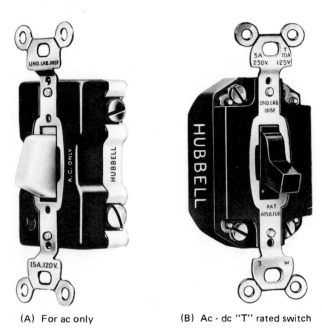

(A) For ac only   (B) Ac - dc "T" rated switch

Fig. 7-1 General-use snap switches.

- motor loads not to exceed 80 percent of the ampere rating of the switch at rated voltage, but not exceeding two horsepower.

Ac general-use snap switches may be marked *ac only*, or they may also be marked with the current and voltage rating markings. A typical switch marking is *15A, 120-277V ac.* The 277-rating is required on 277/480-volt systems. See *Section 210-6* of the *National Electrical Code* for additional information pertaining to maximum voltage limitations.

Terminals of switches rated at 20 amperes or less, when marked *CO/ALR* are suitable for use with aluminum, copper, and copper-clad aluminum conductors. Switches not marked *CO/ALR* are suitable for use with copper, and copper-clad conductors only.

Screwless pressure terminals of the conductor push-in type may be used with copper and copper-clad aluminum conductors only. These push-in type terminals are not suitable for use with ordinary aluminum conductors.

Further information on switch ratings is given in *NEC Section 380-14* and in the Underwriters Laboratories *Electrical Construction Materials List*.

Switch Types and Connections

Switches are readily available in four basic types: single-pole, three-way, four-way and double-pole.

Single-pole Switch. A single-pole switch is used where it is desired to control a light or group of lights, or other load, from one switching point.

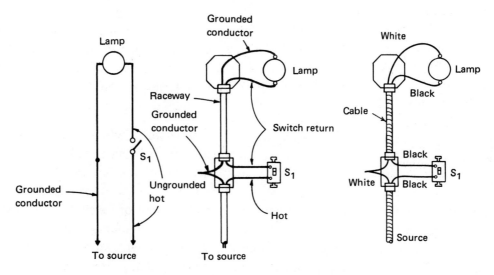

(A) Circuit with single-pole switch feed at switch

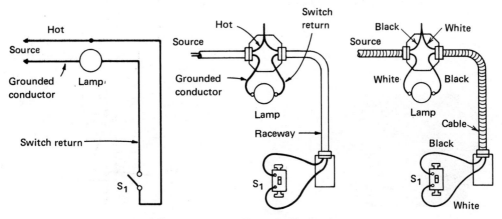

(B) Circuit with single-pole switch feed at light

Fig. 7-2 Single-pole switch connection.

This type of switch is used in series with the ungrounded (hot) wire feeding the load. Figure 7-2 shows typical applications of a single-pole switch controlling a light from one switching point, either at the switch or at the light. Note that the identified, grounded wire goes directly to the load and the unidentified, ungrounded wire is broken at the single-pole switch. *NEC Section 380-2(b)*.

Three-way Switch. A three-way switch can be compared with a single-pole, double-throw switch. There is a *common terminal* to which the switch blade is always connected. The other two terminals are called the *traveler terminals*, figure 7-3. In one position, the switch blade is connected between the common terminal and one of the traveler terminals. In the other position, the switch blade is connected between the common terminal and the second traveler terminal. The three-way switch can be identified readily because it has no *On* or *Off* position. Note that *On* and *Off* positions are not marked on the switch handle in figure 7-3. The three-way switch is also identified by its three terminals. The common terminal is darker in color than the two traveler terminals which have a natural brass color. Figure 7-4 shows the application of three-way switches to provide control at the light or at the switch.

Four-way Switch. A four-way switch is similar to the three-way switch in that it does not have *On* and *Off* positions. However, the four-way switch has four terminals. Two of these terminals are connected to traveler wires from one three-way switch and the other two terminals are connected

to traveler wires from another three-way switch, figure 7-5. In figure 7-5, terminals A1 and A2 are connected to one three-way switch and terminals B1 and B2 are connected to the other three-way switch. In position 1, the switch connects A1 to B1 and A2 to B2. In position 2, the switch connects A1 to B2 and A2 to B1.

The four-way switch is used when a light or a group of lights, or other load, must be controlled from more than two switching points. The switches that are connected to the source and the load are three-way switches. At all other control points, however, four-way switches are used. Figure 7-6 illustrates a typical circuit in which a lamp is controlled from any one of three switching points. Care must be used to ensure that the traveler wires are connected to the proper terminals of the four-way switch. That is, the two traveler wires from one three-way switch must be connected to the two terminals on one end of the four-way switch. Similarly, the two traveler wires from the other three-way switch must be connected to the two terminals on the other end of the four-way switch.

Double-pole Switch. A double-pole switch is rarely used on lighting circuits. As shown in figure 7-7, a double-pole switch can be used for those installations where two separate circuits are to be controlled with one switch. All conductors of circuits supplying gasoline dispensing pumps, or running through such pumps, must have a disconnecting means. Thus, the lighting mounted on gasoline dispensing islands may require two-pole switches, *Section 514-5*.

(A)

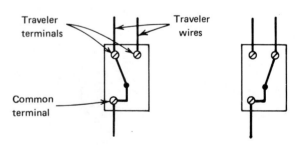

(B) Two positions of a three-way switch

Fig. 7-3 Three-way switch.

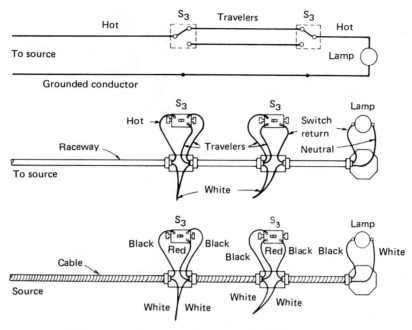

(A) Circuit with three-way switch control and feed at the switch

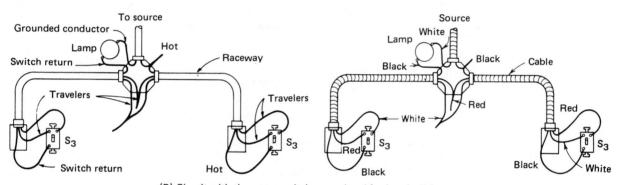

(B) Circuit with three-way switch control and feed at the light

Fig. 7-4 Three-way switch connections.

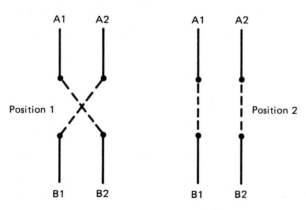

Two positions of four-way switch

Fig. 7-5 Four-way switch operation.

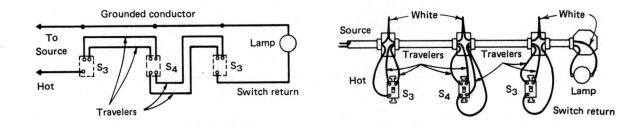

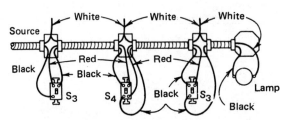

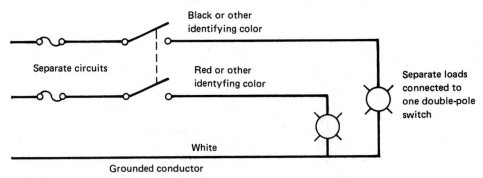

NOTE: A four-way switch is always connected between two three-way switches.

Circuit with switch control at three different locations.

Fig. 7-6 Four-way switch connections.

Fig. 7-7 Double-pole switch connection.

REVIEW

Note: Refer to the *National Electrical Code* or the plans as necessary. (*NEC Article 380*)

Indicate which of the following switches may be used to control the loads listed in questions 1–7.

A. Ac/dc   10 A-125 V   5 A-250 V
B. Ac only   10 A-120 V
C. Ac only   15 A-120/277 V
D. Ac/dc   20 A-125 V-T   10 A-250 V

1. A 120-volt incandescent lamp load (tungsten filament) consisting of ten 150-watt lamps _____

2. A 120-volt fluorescent lamp load (inductive) of 1,500 volt-amperes _____

3. A 277-volt fluorescent lamp load of 625 volt-amperes _____

4. A 120-volt motor drawing 10 amperes _____

5. A 120-volt resistive load of 1,250 watts _____

6. A 120-volt incandescent lamp load of 2,000 watts _____

7. A 230-volt motor drawing 3 amperes _____

8. Show all wiring for the following circuit. The circuit consists of two three-way switches controlling two luminaires with the source at one luminaire. Indicate conductor colors.

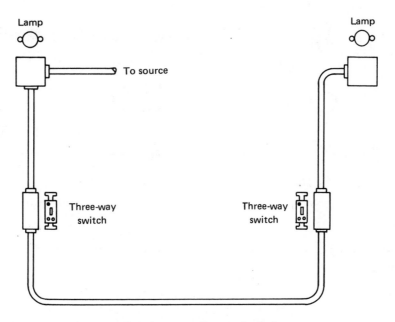

Two three-way switches controlling two luminaires
with source at one luminaire. Show wiring scheme.

9. Show all wiring for the following circuit. The circuit consists of two three-way and one four-way switch controlling a light. Of the three switches, select one as the four-way switch and complete the wiring accordingly. Indicate conductor colors.

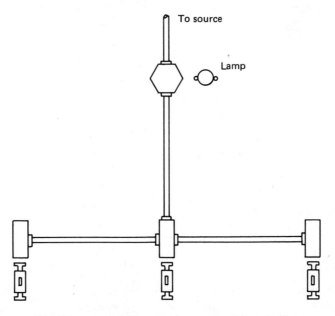

Two three-way switches and a four-way switch controlling
a light. Choose which switch should be the four-way and
show wiring scheme.

UNIT 8

Branch-circuit Installation

OBJECTIVES

After completing the study of this unit, the student will be able to

- complete the installation of rigid conduit and electrical metallic tubing, using the correct materials.

- select the correct conduit size for a branch-circuit installation, using appropriate calculations.

- select the correct size and type of box for the installation, and the conductors to be installed to the box.

- specify the proper raceway support.

- demonstrate the proper method of connecting copper and aluminum conductors.

BRANCH-CIRCUIT INSTALLATION

In a commercial building, the major part of the electrical work is the installation of the branch-circuit wiring. The electrician must have the ability to select and install the correct materials to insure a successful job.

The term *raceway*, which is used in this unit as well as others, is defined by the *NEC* as any channel for holding wires, cables or bus bars that is designed and used expressly for this purpose.

The following paragraphs describe several types of materials which are classified as raceways, including rigid metal conduit, electrical metallic tubing, intermediate metal conduit, flexible metal conduit, and rigid nonmetallic conduit.

RIGID METAL CONDUIT (*NEC Article 346*)

Rigid metal conduit, figure 8-1, is of heavy-wall construction to provide a maximum degree of physical protection to the conductors run through it. Rigid conduit is available in either steel or aluminum. The conduit can be threaded on the job, or nonthreaded fittings may be used where permitted by local codes, figure 8-2.

Rigid conduit bends can be purchased or they can be made using special tools. Bends in 1/2-inch, 3/4-inch and 1-inch conduit can be made using hand benders or hickeys, figure 8-3. Hydraulic benders must be used to make bends in larger sizes of conduit. When necessary, conduit fittings, figure 8-4, or pull boxes, figure 8-5, are used. If pull boxes are used, they must be sized according to *NEC Section 370-18,* as shown in figure 8-6. The following *NEC* requirements apply to the installation of pull boxes and junction boxes.

Fig. 8-1 Rigid steel conduit.

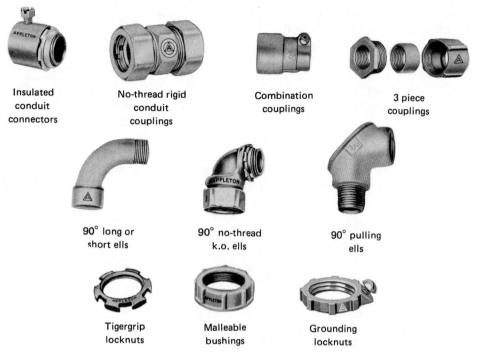

Insulated conduit connectors

No-thread rigid conduit couplings

Combination couplings

3 piece couplings

90° long or short ells

90° no-thread k.o. ells

90° pulling ells

Tigergrip locknuts

Malleable bushings

Grounding locknuts

Fig. 8-2 Rigid metal conduit fittings.

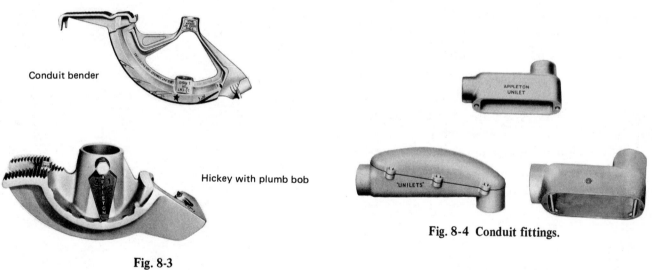

Conduit bender

Hickey with plumb bob

Fig. 8-3

APPLETON UNILET

"UNILETS"

Fig. 8-4 Conduit fittings.

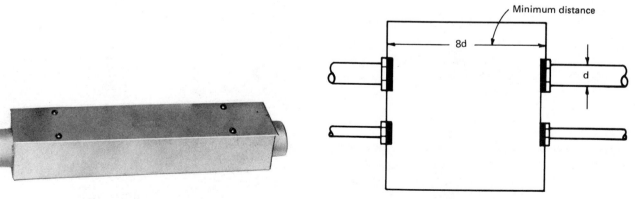

Fig. 8-5 Threaded pull box.

Minimum distance

8d

d

Fig. 8-6 Straight pull. *Section 370-18.*

- For straight pulls, the box must be at least 8 times the diameter of the largest raceway when the raceways are 3/4 inch or larger in diameter and contain No. 3 AWG or larger conductors.

- For angle (U) pulls, figure 8-7, the distance between the raceways and the opposite wall of the box must be at least 6 times the diameter of the largest raceway. To this value, it is necessary to add the diameters of all other raceways entering the same wall of the box. This requirement applies when the raceways are 3/4 inch or larger in diameter and contain No. 4 AWG or larger conductors.

- Boxes may be smaller than the preceding requirements when approved and marked with the size and number of conductors permitted.

- Conductors must be *racked* if any dimension of the box exceeds 6 feet.

- Boxes must have an approved cover.

- Boxes must be accessible.

ELECTRICAL METALLIC TUBING
(NEC Article 348)

The specification for the commercial building in the plans requires the use of electrical metallic tubing (EMT), figure 8-9, as the raceway for all telephone branch circuits and feeder wiring where a conduit size of 2 inches or less is required. *Article 348* of the *NEC* governs the installation of EMT.

EMT Fittings

Electrical metallic tubing is nonthreaded, thin-wall conduit. Since EMT is not to be threaded, conduit sections are joined together and connected to boxes, other fittings, or cabinets by fittings called *couplings* and *connectors*. Several styles of EMT fittings are available, including setscrew, compression, and indenter styles.

Setscrew. When used with this type of fitting, the EMT is pushed into the coupling or connector and is secured in place by tightening the setscrews, figure 8-10. This type of fitting is classified as concrete-tight.

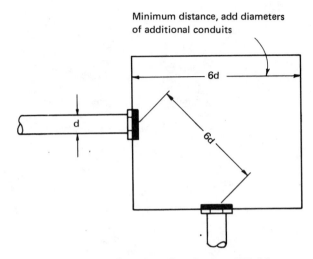

Minimum distance, add diameters of additional conduits

Fig. 8-7 Angle or U pulls. *Section 370-18.*

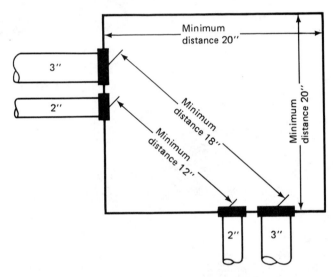

Fig. 8-8 Example of angle pull when more than one raceway enters same wall of box.

Fig. 8-9 Electrical metallic tubing.

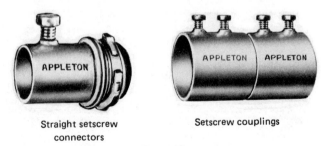

Straight setscrew connectors   Setscrew couplings

Fig. 8-10

Compression. EMT is secured in these fittings by tightening the compressing nuts with a wrench or pliers, figure 8-11. These fittings are classified as raintight and concrete-tight types.

Indenter. A special tool is used to secure EMT in this style of fitting. The tool places an indentation in both the fitting and the conduit. It is a standard wiring practice to make two sets of indentations at each connection. This type of fitting is classified as concrete-tight. Figure 8-12 shows a straight indenter connector.

The efficient installation of EMT requires the use of a bender similar to the one used for rigid conduit, figure 8-13. This tool is commonly

Fig. 8-11

Staight indenter connectors, with or without insulated throat

Fig. 8-12

available in hand-operated models for EMT in sizes from 1/2 inch to 1 inch, and in power-operated models for EMT in sizes greater than 1 inch.

Three kinds of bends can be made with the use of the bending tool. The stub bend, the back-to-back bend, and the angle bend are shown in figure 8-14. The manufacturer's instructions which accompany each bender indicate the method of making each type of bend.

INTERMEDIATE METAL CONDUIT

Intermediate metal conduit has a wall thickness that is between that of rigid metal conduit and EMT. This type of conduit can be installed using either threaded or nonthreaded fittings. *Article 345* of the *NEC* defines intermediate metal conduit, and covers the uses permitted and not permitted, the installation requirements, and the construction specifications for intermediate metal conduit.

INSTALLATION

Rigid metal conduit, intermediate metal conduit, and electrical metallic tubing are installed according to the requirements of *Articles 345, 346,* and *348* of the *NEC.* The following points summarize the contents of these articles.

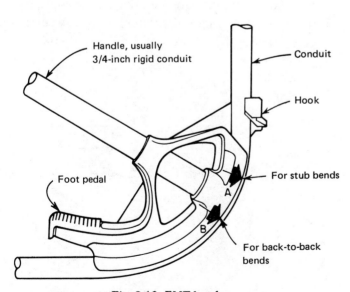

Fig. 8-13 EMT bender.

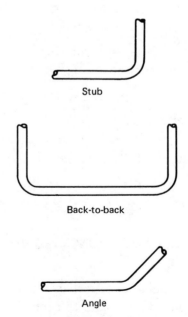

Fig. 8-14 Conduit bends.

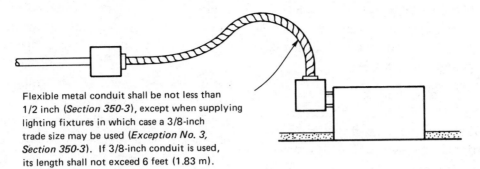

Flexible metal conduit shall be not less than 1/2 inch (*Section 350-3*), except when supplying lighting fixtures in which case a 3/8-inch trade size may be used (*Exception No. 3, Section 350-3*). If 3/8-inch conduit is used, its length shall not exceed 6 feet (1.83 m).

Fig. 8-15 Flexible metal conduit.

The rigid types of metal conduit:

- may be installed in concealed and exposed work.

- may be installed in or under concrete when of the type approved for this purpose.

- must not be installed in or under cinder concrete or cinder fill that is subject to permanent moisture unless the conduit is encased in at least two inches (50.8 mm) of noncinder concrete or is at least 18 inches (457 mm) under the fill, or is of corrosion-resistant material suitable for the purpose.

- must not be installed where subject to severe mechanical damage (an exception to this is rigid metal conduit which may be installed in a location where it is subject to damaging conditions).

- may contain up to four quarter bends (for a total of 360 degrees) in any run.

- must be fastened within three feet (914 mm) of each outlet box.

- must be supported at least every 10 feet (3.05 m).

- may be installed in wet or dry locations if the conduit is of the type approved for this use.

- must have the ends reamed to remove rough edges.

FLEXIBLE CONNECTIONS (*NEC Articles 350* and *351*)

The installation of certain equipment requires flexible connections, both to simplify the installation and to stop the transfer of vibrations.

The two basic types of material used for these connections are flexible metal conduit, figure 8-15, and liquidtight flexible metal conduit, figure 8-16.

Article 350 of the Code covers the use and installation of flexible metal conduit. This type of conduit is similar to armored cable, except that the conductors are installed by the electrician. For armored cable, the cable armor is wrapped around the conductors at the factory to form a complete cable assembly.

Some of the more common installations using flexible metal conduit are shown in figure 8-17. Note that the flexibility required to make the installation is provided by the flexible metal conduit. The figure calls attention to the *National Electrical Code* and Underwriters Laboratories restrictions on the use of flexible metal conduit with regard to relying on the metal armor as a grounding means.

The use and installation of liquidtight flexible metal conduit are described in *Article 351* of the Code. Liquidtight flexible metal conduit has a tighter fit of its spiral turns as compared

Fig. 8-16 Liquidtight flexible conduit and fitting.

to standard flexible metal conduit. Liquidtight conduit also has a thermoplastic outer jacket that is liquidtight. Liquidtight flexible metal conduit is commonly used as the flexible connection to a central air-conditioning unit located outdoors, figure 8-18.

Figure 8-19 shows the limitations placed on the use of liquidtight flexible metal conduit as a grounding means. These limitations are given in the Underwriters Laboratories Standards.

Liquidtight flexible *nonmetallic* conduit may be used:

- in industrial applications only.
- where flexibility is required.
- where not subject to physical damage.

- where temperatures do not exceed that for which the flexible conduit is approved.
- in lengths not over 6 feet (1.83 m).
- with fittings identified for use with the flexible conduit.
- in sizes 1/2 inch to 2 inches. For enclosing motor leads, 3/8 inch is suitable.

When fittings are not marked *GRND*, it can be assumed that they are not approved for equipment grounding purposes. In this case, a separate equipment grounding conductor must be installed. The conductor is sized according to *NEC Table 250-95*. Figure 8-20 illustrates the application of this table.

A FLEXIBLE METAL CONDUIT MAY BE USED IN TRADE SIZES NOT OVER 3/4 INCH AS A GROUNDING MEANS IF IT IS NOT OVER 6 FEET (1.83 m) LONG, IS CONNECTED BY FITTINGS APPROVED FOR GROUNDING PURPOSES, AND THE CIRCUIT OVERCURRENT DEVICE IS RATED AT NOT OVER 20 AMPERES. SEE *NEC ARTICLE 350* AND THE UNDERWRITERS LABORATORIES *GREEN BOOK*.

B FLEXIBLE METAL CONDUIT IN TRADE SIZES LARGER THAN 3/4 INCH MAY BE USED FOR GROUNDING IF IT IS NOT LONGER THAN SIX FEET (1.83 m) BUT ONLY WHEN THE FITTINGS ARE MARKED "GRND". FLEXIBLE METAL CONDUIT OF ANY SIZE, WHEN LONGER THAN SIX FEET MAY *NOT* BE USED AS A GROUNDING MEANS (SEE UNDERWRITERS LABORATORIES *GREEN BOOK*.) IN THIS CASE, A BONDING JUMPER IS REQUIRED.

C WHEN REQUIRED BY THE CODE, THE BONDING JUMPER MAY BE INSTALLED INSIDE OR OUTSIDE THE FLEXIBLE METAL CONDUIT. THE BONDING JUMPER CAN BE INSTALLED OUTSIDE THE CONDUIT ONLY WHEN THE CONDUIT IS NOT OVER SIX FEET (1.83 m) LONG.

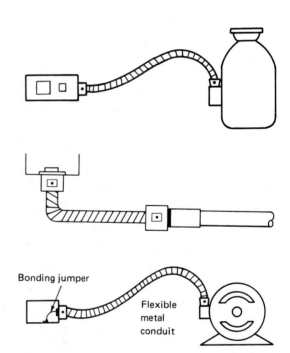

Fig. 8-17 Installations using flexible metal conduit.

LIQUIDTIGHT FLEXIBLE METAL CONDUIT MAY BE USED AS A GROUNDING MEANS IF IT IS NOT OVER 1 1/4-INCH TRADE SIZE, IS NOT OVER SIX FEET (1.83 m) LONG, AND IS CONNECTED BY FITTINGS APPROVED FOR GROUNDING PURPOSES.

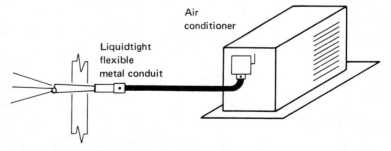

Fig. 8-18 A use of liquidtight flexible metal conduit.

LIQUIDTIGHT FLEXIBLE METAL CONDUIT MAY NOT BE USED AS A GROUNDING MEANS IN ANY SIZE WHEN LONGER THAN SIX FEET (1.83 m) OR IN SIZES 1 1/2 INCH AND LARGER OF ANY LENGTH.   IN SUCH CASES, A BONDING JUMPER IS REQUIRED.

Fig. 8-19 Limitations on the use of liquidtight flexible metal conduit.

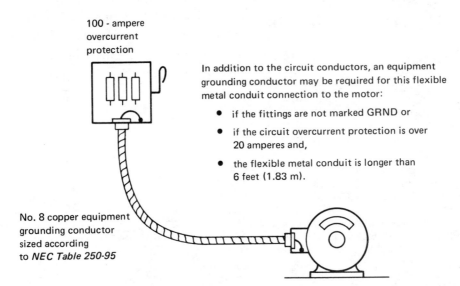

100 - ampere overcurrent protection

In addition to the circuit conductors, an equipment grounding conductor may be required for this flexible metal conduit connection to the motor:

• if the fittings are not marked GRND or

• if the circuit overcurrent protection is over 20 amperes and,

• the flexible metal conduit is longer than 6 feet (1.83 m).

No. 8 copper equipment grounding conductor sized according to *NEC Table 250-95*

Fig. 8-20 Separate equipment grounding conductor in flexible metal conduit installation.

RIGID NONMETALLIC CONDUIT

Rigid nonmetallic conduit is made of a synthetic material called polyvinyl chloride (PVC). Threaded or slip-on fittings may be used with this type of conduit, which is recommended for installations in wet locations, in certain corrosive atmospheres, and in cinder fills. It is very important that expansion joints be installed on this type of conduit as recommended by the manufacturer. The electrician must also remember that a grounding conductor must be installed and connected to all of the metallic junction boxes used in the installation. *NEC Article 347* provides additional information for the installation of rigid nonmetallic conduit.

CONDUIT SIZING

The conduit size required for an installation depends upon three factors: (1) the number of conductors to be installed, (2) the cross-sectional area of the conductor, and (3) the permissible conduit fill. The relationship of these factors is defined in *Chapter 9* of the *NEC* in the *Notes* and *Tables 1, 2, 3A, 3B,* and *3C.* After the student examines the plans and determines the number of conductors to be installed to a certain point, either of the following procedures can be used to find the conduit size.

If all of the conductors have the same insulation, the conduit size can be determined directly by referring to *Tables 3A, 3B,* or *3C* of *Chapter 9.* For example, three No. 8 THWN conductors may be installed to the air-conditioning unit in the drugstore in the commercial building. Refer to *Table 3A* and find the type letter (THWN) and the conductor size (8). When these items are located, it can be seen that 1/2-inch conduit is allowed.

Now assume that in addition to the three No. 8 THWN conductors, a No. 10 THWN equipment grounding conductor is to be installed. This type of installation will meet the requirements of *Article 250* if a flexible connection is used to connect to the air-conditioning unit. The No. 10 grounding conductor must be included in any determination of the required conduit size as specified by *Chapter 9, Note 2*.

Thus, for the case of three No. 8 THWN conductors and one No. 10 THWN grounding conductor, refer to *Table 5* to determine the conductor area. The total cross-sectional area of all of the conductors can now be calculated.

1 No. 10 THWN	0.0184 sq. in.
3 No. 8 THWN @ 0.0373 sq. in.	0.1119 sq. in.
Total conductor cross-sectional area	0.1303 sq. in.

Next, refer to *Table 4* under the heading "Not Lead-covered" and the subheading "Over 2 Cond. 40%" to find the next larger value of area. Follow the line on which the area is located to the left to find the conduit size. For this situation, 3/4-inch conduit is required.

BOX STYLES AND SIZING

The style of box required on a building project is usually established in the specifications. However, the sizing of the boxes is usually one of the decisions made by the electrician.

Switchboxes

Switchboxes are 2" x 3" in size and are available with depths ranging from 1 1/2 inches to 3 1/2 inches, figure 8-21. These boxes can be purchased for either 1/2-inch or 3/4-inch conduit. Each side of the switchbox has holes through which a 20-penny nail can be inserted for nailing to wood studs. The boxes can be ganged by removing the common sides of two or more boxes and connecting the boxes together. Plaster ears may be provided for use on plasterboard or for work on old installations.

Masonry Boxes

Masonry boxes are designed for use in masonry block or brick. These boxes do not require an extension cover, but will accommodate devices directly, figure 8-22. Masonry boxes are available

with depths of 2 1/2 inches and 3 1/2 inches and in through-the-wall depths of 3 1/2 inches, 5 1/2 inches, and 7 1/2 inches. Boxes of this type are available with knockouts up to one inch in diameter.

Handy Box

Handy boxes, figure 8-23, are generally used in exposed installations and are available with 1/2-, 3/4- or 1-inch knockouts. Since these boxes range in depth from 1 1/4 inches to 2 1/2 inches, they can accommodate a device without the use of an extension cover.

3 x 2 inch conduit switch boxes in assorted depths with or without plaster ears

Fig. 8-21

1, 2, 3, 4 or 5-gang masonry boxes

Fig. 8-22

Handy box series is available in different knockout combinations and with front or side eagle claw mounting brackets

Assorted covers for handy boxes are available for switches and receptacles; blank covers are available also

Fig. 8-23

4-inch Square Boxes

A 4-inch square box, figure 8-24, is used commonly for surface or concealed installations. Extension covers of various depths are available to accommodate devices in those situations where the box is surface mounted. This type of box is available with knockouts up to one inch in diameter.

Octagonal Boxes

Boxes of this type are used primarily to install ceiling outlets. Octagonal boxes are available either for mounting in concrete or for surface or concealed mounting, figure 8-25. Extension covers are available, but are not always required. Octagonal boxes are commonly used in depths of 1 1/2 inches and 2 1/8 inches. These boxes are available with knockouts up to an inch in trade diameter.

4 11/16-inch Boxes

These more spacious boxes are used where the larger size is required. Available with 1/2-, 3/4-, 1- and 1 1/4-inch knockouts, these boxes require an extension cover or a raised cover to permit the attachment of devices, figure 8-26.

Box Sizing

After selecting the style of box to be used, the electrician must determine the correct box size according to the requirements of *NEC Section 370-6*, figure 8-27. An example of the procedure for selecting the box size is the installation of a four-way switch in the bakery. Four No. 12 AWG conductors and one device are to be installed in the box. Therefore, the box selected must be able to accommodate five conductors; a 3″ x 2″ x 2 1/2″ switch box is adequate. Another box is needed where the conduit from a four-way switch terminates at the lighting outlet. The items entering the box consist of four No. 12 travelers, two No. 12 switch returns, two No. 12 neutrals, and one fixture stud-hickey-stem assembly. Since the box must be able to accommodate nine No. 12 conductors, a 4-inch octagonal box, 2 1/8 inches deep will be required.

Octagonal box with bar hanger

Octagonal concrete box

Fig. 8-25

4-inch square drawn outlet box with eagle claw mounting brackets, cable clamps and conduit knockouts

4-inch square extension ring in assorted knockout combinations

Raised 4-inch square cover available for toggle switches and receptacles

4-inch square single offset device

Fig. 8-24

4 11/16-inch square outlet box series for conduit is available in various depth and knockout combinations. Extension rings are also listed

Fig. 8-26

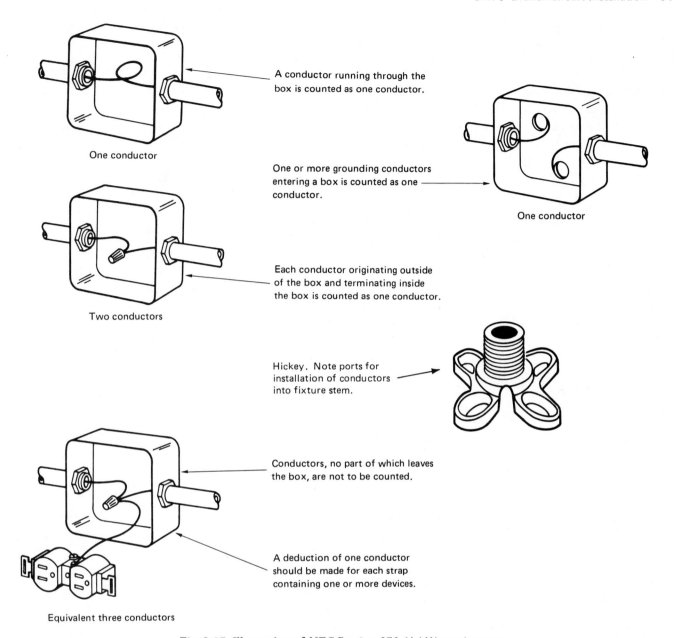

One conductor

A conductor running through the box is counted as one conductor.

One or more grounding conductors entering a box is counted as one conductor.

One conductor

Two conductors

Each conductor originating outside of the box and terminating inside the box is counted as one conductor.

Hickey. Note ports for installation of conductors into fixture stem.

Conductors, no part of which leaves the box, are not to be counted.

A deduction of one conductor should be made for each strap containing one or more devices.

Equivalent three conductors

Fig. 8-27 Illustration of *NEC Section 370-6(a)(1)* requirements.

When the box contains devices, such as fixture studs, clamps, or hickeys, the number of conductors permitted in the box is one less for each type of device. For example, if a box contains two clamps, deduct one conductor from the allowable amount as shown in *Table 370-6(a)*.

RACEWAY SUPPORT

There are many devices available which are used to support conduit. The more popular devices are shown in figure 8-28.

The *NEC* article which deals with a particular type of raceway also gives the requirements for supporting that raceway. Figure 8-29 shows the support requirements for rigid metal conduit, intermediate metal conduit, electrical metallic tubing, flexible metal conduit, and liquidtight flexible metal conduit.

ALUMINUM CONDUCTORS

The conductivity of aluminum is not as great as that of copper for a given wire size. For example, checking *NEC Table 310-16,* the ampacity for copper No. 12 AWG Type THW is 25 amperes, whereas the ampacity for aluminum No. 12 AWG Type THW is 20 amperes. As another example, a No. 8 TW copper wire has an ampacity of 40 amperes, whereas a No. 8 TW aluminum or copper-clad aluminum wire has an ampacity of only 30 amperes. It is important to check the ampacities listed in *NEC Table 310-16* and the footnote to the table.

Resistance is an important consideration when installing aluminum conductors. An aluminum conductor has a higher resistance compared to a copper conductor for a given wire size which, therefore, causes a greater voltage drop.

$$\text{Voltage Drop } (E_d) = \text{Amperes } (I) \times \text{Resistance } (R)$$

Table 370-6(a).   Metal Boxes

Box Dimension, Inches Trade Size or Type	Min. Cu. In. Cap.	Maximum Number of Conductors				
		No. 14	No. 12	No. 10	No. 8	No. 6
4 x 1¼ Round or Octagonal	12.5	6	5	5	4	0
4 x 1½ Round or Octagonal	15.5	7	6	6	5	0
4 x 2⅛ Round or Octagonal	21.5	10	9	8	7	0
4 x 1¼ Square	18.0	9	8	7	6	0
4 x 1½ Square	21.0	10	9	8	7	0
4 x 2⅛ Square	30.3	15	13	12	10	6*
4¹¹⁄₁₆ x 1¼ Square	25.5	12	11	10	8	0
4¹¹⁄₁₆ x 1½ Square	29.5	14	13	11	9	0
4¹¹⁄₁₆ x 2⅛ Square	42.0	21	18	16	14	6
3 x 2 x 1½ Device	7.5	3	3	3	2	0
3 x 2 x 2 Device	10.0	5	4	4	3	0
3 x 2 x 2¼ Device	10.5	5	4	4	3	0
3 x 2 x 2½ Device	12.5	6	5	5	4	0
3 x 2 x 2¾ Device	14.0	7	6	5	4	0
3 x 2 x 3½ Device	18.0	9	8	7	6	0
4 x 2⅛ x 1½ Device	10.3	5	4	4	3	0
4 x 2⅛ x 1⅞ Device	13.0	6	5	5	4	0
4 x 2⅛ x 2⅛ Device	14.5	7	6	5	4	0
3¾ x 2 x 2½ Masonry Box/Gang	14.0	7	6	5	4	0
3¾ x 2 x 3½ Masonry Box/Gang	21.0	10	9	8	7	0
FS—Minimum Internal Depth 1¾ Single Cover/Gang	13.5	6	6	5	4	0
FD—Minimum Internal Depth 2⅜ Single Cover/Gang	18.0	9	8	7	6	3
FS—Minimum Internal Depth 1¾ Multiple Cover/Gang	18.0	9	8	7	6	0
FD—Minimum Internal Depth 2⅜ Multiple Cover/Gang	24.0	12	10	9	8	4

* Not to be used as a pull box. For termination only.

Common Connection Problems

Some common problems associated with aluminum conductors when not properly connected may be summarized as follows:

- a corrosive action is set up when dissimilar wires come in contact with one another when moisture is present.

- the surface of aluminum oxidizes as soon as it is exposed to air. If this oxidized surface is not broken through, a poor connection results. When installing aluminum conductors, particularly in large sizes, an inhibitor is brushed onto the aluminum conductor, then the conductor is scraped with a stiff brush where the connection is to be made. The process of scraping the conductor breaks through the oxidation, and the inhibitor keeps the air from coming into contact with the conductor. Thus, further oxidation is prevented. Aluminum connectors of the compresssion type usually have an inhibitor paste already factory installed inside the connector.

Table 370-6(b).   Volume Required Per Conductor

Size of Conductor	Free Space Within Box for Each Conductor
No. 14	2. cubic inches
No. 12	2.25 cubic inches
No. 10	2.5 cubic inches
No. 8	3. cubic inches
No. 6	5. cubic inches

Reprinted with permission from NFPA 70-1984, *National Electrical Code®*, Copyright © 1983, National Fire Protection Association, Quincy, Massachusetts 02269. This reprinted material is not the complete and official position of the NFPA on the referenced subject, which is represented only by the standard in its entirety.

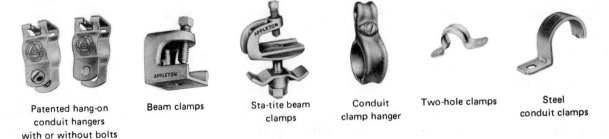

Patented hang-on conduit hangers with or without bolts   Beam clamps   Sta-tite beam clamps   Conduit clamp hanger   Two-hole clamps   Steel conduit clamps

Fig. 8-28 Raceway support devices.

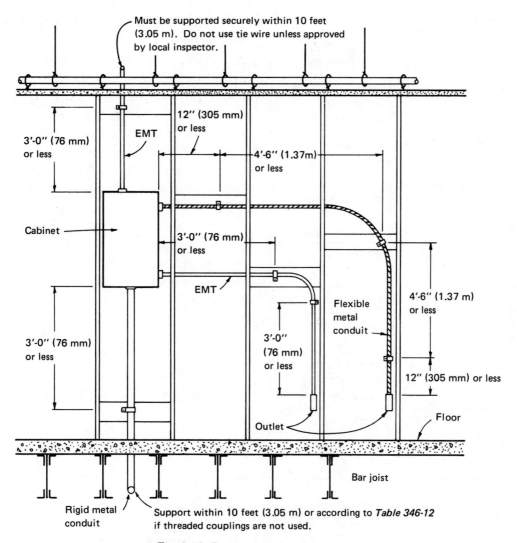

Must be supported securely within 10 feet (3.05 m). Do not use tie wire unless approved by local inspector.

3'-0'' (76 mm) or less

EMT

12'' (305 mm) or less

4'-6'' (1.37m) or less

Cabinet

3'-0'' (76 mm) or less

EMT

Flexible metal conduit

4'-6'' (1.37 m) or less

3'-0'' (76 mm) or less

3'-0'' (76 mm) or less

12'' (305 mm) or less

Floor

Outlet

Bar joist

Rigid metal conduit

Support within 10 feet (3.05 m) or according to *Table 346-12* if threaded couplings are not used.

Fig. 8-29 Supporting raceway.

- aluminum wire expands and contracts to a greater degree than does copper wire for an equal load. This factor is another possible cause of a poor connection. Crimp connectors for aluminum conductors are usually longer than those for comparable copper conductors, thus resulting in greater contact surface of the conductor in the connector.

PROPER INSTALLATION PROCEDURES

Proper, trouble-free connections for aluminum conductors require terminals, lugs, and/or connectors which are suitable for the type of conductor being installed.

Records indicate that the majority of failures in the use of aluminum wire have occurred because of poor installation. The electrician should take special care to make sure that all connections are thoroughly tight. Such devices as "Yankee screwdrivers" should not be used for installation of terminals, because these tools often give a false impression as to the tightness of the connection.

Terminals on receptacles and switches must be suitable for the conductors being attached. The chart on this page shows how the electrician can identify these terminals.

Wire Connections

When splicing wires or connecting a wire to a switch, fixture, circuit breaker, panelboard, meter socket or other electrical equipment, the wires may be twisted together, soldered, then taped. Usually, however, some type of wire connector is required.

Ampacity	Marking on Terminal or Connector	Conductor Permitted
15- or 20 ampere receptacles and switches	CO/ALR	aluminum, copper, copper-clad aluminum
15- and 20-ampere receptacles and switches	NONE	copper, copper-clad aluminum
30-ampere and greater receptacles and switches	AL/CU	aluminum, copper, copper-clad aluminum
30-ampere and greater receptacles and switches	NONE	copper only
Screwless pressure terminal connectors of the push-in type	NONE	copper or copper-clad aluminum
Wire connectors	AL/CU	aluminum, copper, copper-clad aluminum
Wire connectors	NONE	copper only
Wire connectors	AL	aluminum only

Wire connectors are known in the trade by such names as *screw terminal, pressure terminal connector, wire connector, wing nut, wire nut, Scotchlok, split-bolt connector, pressure cable connector, solderless lug, soldering lug, solder lug,* and others. Soldering-type lugs are not often used today. Solderless connectors, designed to establish connections by means of mechanical pressure, are quite common. Examples of some types of wire connectors, and their uses, are shown in figure 8-30.

As with the terminals on wiring devices (switches and receptacles) wire connectors must be marked *AL* when they are to be used with aluminum conductors. This marking is found on the connector itself, or it appears on or in the shipping carton.

Connectors marked *AL/CU* are suitable for use with aluminum, copper, or copper-clad aluminum conductors. This marking is found on the connector itself, or it appears on or in the shipping carton.

Connectors not marked *AL* or *AL/CU* are for use with copper conductors only.

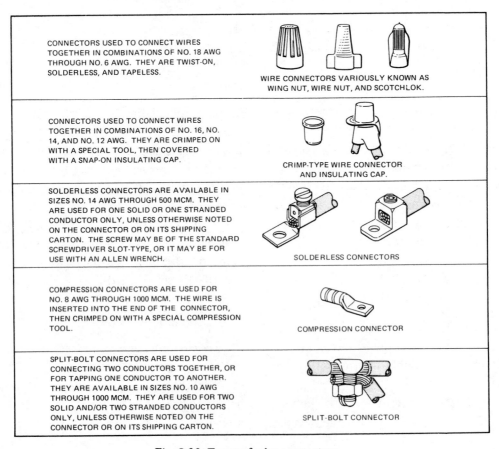

Fig. 8-30 Types of wire connectors.

Unless specially stated on or in the shipping carton, or on the connector itself, conductors made of copper, aluminum, or copper-clad aluminum may not be used in combination in the same connector. Combinations, when permitted, are usually limited to dry locations only.

REVIEW

Note: Refer to the *National Electrical Code* or the plans as necessary.

1. a. For the following diagram, the minimum acceptable length dimension is _____ inches (_____ millimeters).
   b. The minimum acceptable width dimension is _____ inches (_____ millimeters).

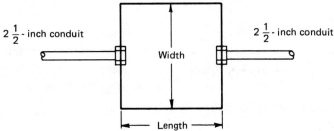

2. a. For the following figure, dimension *a* must be at least _____ inches (_____ millimeters).
   b. Dimension *b* must be at least _____ inches (_____ millimeters).
   c. Dimension *c* must be at least _____ inches (_____ millimeters).

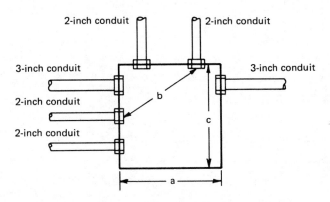

3. Four No. 12 AWG and four No. 10 AWG conductors are to be installed in a conduit. What size conduit is required if:
   a. the conductors are type THW? _____

   _____

   b. the conductors are type TW? _____

   _____

4. What is the smallest square box that will accommodate four No. 12 conductors and four No. 10 conductors if:
   a. these conductors are pulled through without a splice? _____

   _____

   b. these conductors are all spliced in the box? _____

   _____

5. For each of the following raceways, how many supports are required to be installed? Both ends of each raceway are terminated in outlet boxes. Threaded couplings will be used with rigid metal conduit.

a. 50 feet (15.2 m) of 1/2-inch EMT _____

_____

b. 18 feet (5.49 m) of 1/2-inch EMT _____

_____

c. 50 feet (15.2 m) of 1-inch rigid metal conduit _____

_____

d. 18 feet (5.49 m) of 1-inch rigid metal conduit _____

_____

6. The allowable current-carrying capacity (ampacity), or the maximum permitted load current in the case of smaller conductors, for aluminum wire is less than that of copper wire for a given size, insulation, and temperature of 86°F. Refer to *Table 310-16* and footnote and complete the following table:

WIRE	COPPER AMPACITY	ALUMINUM AMPACITY
No. 12 TW		
No. 10 TW		
No. 3 THW		
0000 THWN		
500 MCM THWN		

7. It is permissible for an electrician to connect aluminum, copper, or copper-clad aluminum conductors together in the same connector.
True _____   False _____

8. Terminals of switches and receptacles marked *CO/ALR* are suitable for use with _____ , _____ , and _____ conductors.

9. Wire connectors marked *AL/CU* are suitable for use with _____ , _____ , and _____ conductors.

10. A wire connector bearing no marking or reference to *AL, CU,* or *ALR* is suitable for use with (copper) (aluminum) conductors only. (Underline the correct answer.)

UNIT 9

Appliance Circuits

OBJECTIVES

After completing the study of this unit, the student will be able to

- define the meaning of *appliance*.
- identify three types of appliances specified by the *NEC*.
- apply the *NEC* to the installation of appliances.
- select branch-circuit and overload protection for appliances.

APPLIANCES (*NEC Article 100*)

The *National Electrical Code* defines *appliance* as utilization equipment, generally other than industrial, of standardized sizes or types, which is installed or connected as a unit to perform such functions as air conditioning, washing, food mixing, and cooking, among others.

Past issues of the Code attempted to further define an appliance as being "fixed," "portable," or "stationary." Because confusion resulted, the Code-making panels have included the following terms in various sections throughout the Code, where necessary for clarification:

- "permanently connected"
- "cord- and plug-connected"
- "fastened in place"
- "located to be on a specific circuit"

One of the common tasks performed by the electrician is the installation of appliance circuits. Therefore, the electrician should be familiar with the *NEC* sections governing appliance-type equipment. The bakery installation in the commercial building provides examples of several different appliance connection situations.

THE EXHAUST FAN

The exhaust fan is a permanently connected, fastened in place, motor-operated appliance supplied by a general-purpose branch circuit. *NEC Sections 430-42, 210-23*, and *210-24* apply to this installation. The exhaust fan has a 1/10-hp, 120-volt motor with an FLA rating of 2.9 amperes. Since the motor has a rating of less than 10 amperes, it may be connected to a 20-ampere branch circuit. An ac general-use snap switch is installed under the fan to provide control of the fan. A handy box is located next to the fan to permit the use of flexible conduit to the motor, figure 9-1.

If the installation is permanent, then overload protection is required for the 120-volt motor since it is connected to a 20-ampere branch circuit. The overload protection, in this case, consists of a dual-element plug fuse. This type of fuse is designed to withstand the motor starting inrush current. However, the fuse will open if a sustained overload occurs. In general, dual-element fuses are sized at not over 125% of the motor full-load rating. Therefore, for the 2.9-ampere rating of the 120-volt exhaust fan motor, 2.9 X 1.25 = 3.62 amperes maximum. Thus, a 3 2/10-ampere time-delay fuse is selected.

The typical one-time fuse contains a copper or zinc fuse link which melts or opens quickly when an overload or short circuit occurs. The dual-element fuse has a copper or silver link as well as a thermal cutout element which opens under continuing overloads. The overload elements are held together by a solder alloy which melts if an overload continues. When the solder melts, a spring pulls the elements apart to open the circuit. In the event that a short circuit occurs and there is a large overcurrent of 500% or more, the copper link in the dual-element fuse melts and also opens the circuit. Table 9-1 compares the load responses of ordinary one-time fuses and dual-element fuses. This table was developed from time-current characteristic curves of the type shown in figure 18-20.

The starting current of a motor with a full-load rating of 4 amperes may be as high as 24 amperes. In this situation, a 12- or 15-ampere ordinary fuse may be required; however, this fuse will allow continuing overloads which can destroy the motor. (See *NEC Table 430-152*). A 4-ampere, dual-element fuse permits the motor to start, yet opens on an overload before the motor is damaged. The Edison-base fuse is the most frequently encountered plug fuse in existing installations, but the Type S plug fuse and adapter, figure 9-2, are required for new installations. (Refer to *NEC Sections 240-53* and *240-54*.) The Type S fuse adapter is used in the Edison-type fuseholder to restrict the size of fuses that can be installed. Fuseholders are available in box cover units such as those shown in figure 9-3.

THE CAKE MIXERS AND THE DOUGH DIVIDER

A separate appliance branch circuit supplies three receptacle outlets on the south wall of the bakery. The three appliances supplied by this circuit are a cake mixer with a full-load rating of

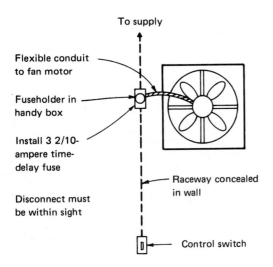

To supply

Flexible conduit to fan motor

Fuseholder in handy box

Install 3 2/10-ampere time-delay fuse

Disconnect must be within sight

Raceway concealed in wall

Control switch

Fig. 9-1 Exhaust fan installation.

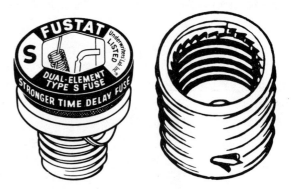

Fig. 9-2 Type S plug fuse and adapter.

TABLE 9-1 LOAD RESPONSE OF FUSES
(Approximate Time in Seconds for Fuse to Blow)

Load (amperes)	4-ampere Dual-element Fuse (time-delay)	4-ampere Ordinary Fuse (nontime-delay)	8-ampere Ordinary Fuse (nontime-delay)	15-ampere Ordinary Fuse (nontime-delay)
5	Over 300 sec.	Over 300 sec.	Won't blow	Won't blow
6	250 sec.	5 sec.	Won't blow	Won't blow
8	60 sec.	1 sec.	Won't blow	Won't blow
10	38 sec.	Less than 1 sec.	Over 300 sec.	Won't blow
15	17 sec.	Less than 1 sec.	5 sec.	Won't blow
20	9 sec.	Less than 1 sec.	Less than 1 sec.	300 sec.
25	5 sec.	Less than 1 sec.	Less than 1 sec.	10 sec.
30	2 sec.	Less than 1 sec.	Less than 1 sec.	4 sec.

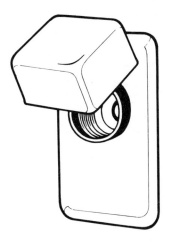

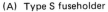

(A) Type S fuseholder

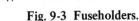

(B) Type S fuseholder and switch

(C) Type S fuseholder with switch and pilot light

Fig. 9-3 Fuseholders.

3.96 amperes, a second cake mixer with a full-load rating of 7.48 amperes, figure 9-4, and a dough divider with a rating of 2.2 amperes, figure 9-5. As specified by the *NEC,* 15- and 20-ampere branch circuits are installed to supply lighting units, appliances, or a combination of both. For such a branch circuit, the rating of any one cord-and plug-connected appliance load shall not exceed 80% of the branch-circuit rating. In addition, the total rating of the appliances fastened in place shall not exceed 50% of the branch-circuit rating when other loads are also supplied. [See *NEC Section 210-23(a)*]. The rating of the fuse selected as the overcurrent protection depends upon the value of the equipment and how critical the shutdown of the equipment is to the overall operation. Maximum circuit protection is obtained by selecting a fuse from Table II of figure 9-6. If a fuse with a higher rating is selected (up to the value indicated in Table I of figure 9-6), less protection is provided, but the equipment will not be shut down as often. See *Table 430-152* of the *NEC* for motor branch-circuit ratings and settings. (Fustat, Fusetron and Low-Peak are copyrighted trade names of the Bussmann Division of the McGraw-Edison Company for Type S or Type T dual-element fuses.)

Abnormal installations may require dual-element, time-delay fuses larger than those shown in Tables I and II of figure 9-6. These larger fuses will provide short-circuit protection only for the motor branch circuit. *NEC Table 430-152* shows the maximum fuse sizes permitted by the Code.

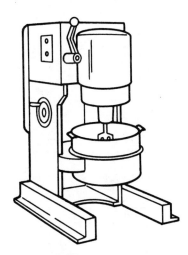

Fig. 9-4 Cake mixer.

Fig. 9-5 Dough divider.

Tables For Selection of Dual-Element Fuses For Motor Running
Overload Protection Based on Motor Nameplate Ampere Rating

TABLE I MOTORS MARKED WITH
NOT LESS THAN 1.15 SERVICE FACTOR OR TEMP. RISE NOT OVER 40° C.

Motor (40° C or 1.15 S.F.) Ampere Rating	Dual-element Fuse Amp Rating * * (Max. 125%; 430-32)	Motor (40° C or 1.15 S.F.) Ampere Rating	Dual-element Fuse Amp Rating * * (Max. 125%; 430-32)
1.00 to 1.11	1 1/4	20.0 to 23.9	25
1.12 to 1.27	1 4/10	24.0 to 27.9*	30
1.28 to 1.43	1 6/10	28.0 to 31.9	35
1.44 to 1.59	1 8/10	32.0 to 35.9	40
1.60 to 1.79	2	36.0 to 39.9	45
1.80 to 1.99	2 1/4	40.0 to 47.9	50
2.00 to 2.23	2 1/2	48.0 to 55.9	60
2.24 to 2.55	2 8/10	56.0 to 63.9	70
2.56 to 2.79	3 2/10	64.0 to 71.9	80
2.80 to 3.19	3 1/2	72.0 to 79.9	90
3.20 to 3.59	4	80.0 to 87.9*	100
3.60 to 3.99	4 1/2	88.0 to 99.9	110
4.00 to 4.47	5	100 to 119	125
4.48 to 4.99	5 6/10	120 to 139	150
5.00 to 5.59	6 1/4	140 to 159	175
5.60 to 6.39	7	160 to 179*	200
6.40 to 7.19	8	180 to 199	225
7.20 to 7.99	9	200 to 239	250
8.00 to 9.59	10	240 to 279	300
9.60 to 11.9	12	280 to 319	350
12.0 to 13.9	15	320 to 359*	400
14.0 to 15.9	17 1/2	360 to 399	450
16.0 to 19.9	20	400 to 480	500

TABLE II ALL OTHER MOTORS
(i.e. LESS THAN 1.15 SERVICE FACTOR OR GREATER THAN 40° C RISE)

All other Motors-Ampere Rating	Dual-element Fuse Amp Rating * * (Max. 115%; 430-32)	All other Motors-Ampere Rating	Dual-element Fuse Amp Rating * * (Max. 115%; 430-32)
1.00 to 1.08	1 1/8	17.4 to 20.0	20
1.09 to 1.21	1 1/4	21.8 to 25.0	25
1.22 to 1.39	1 4/10	26.1 to 30.0*	30
1.40 to 1.56	1 6/10	30.5 to 34.7	35
1.57 to 1.73	1 8/10	34.8 to 39.1	40
1.74 to 1.95	2	39.2 to 43.4	45
1.96 to 2.17	2 1/4	43.5 to 50.0	50
2.18 to 2.43	2 1/2	52.2 to 60.0	60
2.44 to 2.78	2 8/10	60.9 to 69.5	70
2.79 to 3.04	3 2/10	69.6 to 78.2	80
3.05 to 3.47	3 1/2	78.3 to 86.9	90
3.48 to 3.91	4	87.0 to 95.6*	100
3.92 to 4.34	4 1/2	95.7 to 108	110
4.35 to 4.86	5	109 to 125	125
4.87 to 5.43	5 6/10	131 to 150	150
5.44 to 6.08	6 1/4	153 to 173	175
6.09 to 6.95	7	174 to 195*	200
6.96 to 7.82	8	196 to 217	225
7.83 to 8.69	9	218 to 250	250
8.70 to 10.0	10	261 to 300	300
10.5 to 12.0	12	305 to 347	350
13.1 to 15.0	15	348 to 391*	400
15.3 to 17.3	17 1/2	392 to 434	450
		435 to 480	500

*Note: Disconnect switch must have an ampere rating at least 115% of motor ampere rating
(430-110a). Next larger size switch with fuse reducers may be required.

**Use FUSETRON Dual-Element Fuses or LOW-PEAK Dual-Element Fuses
FRN or FRN-R (250 volt)   LPN or LPN-R (250 volt)
FRS or FRS-R (600 volt)   LPS or LPS-R (600 volt)

Fig. 9-6.

Abnormal installations include the following:

- situations where a motor is started, stopped, jogged, inched, or reversed frequently.

- high inertia loads such as large fans and centrifugal machines such as extractors, separators, and pulverizers; in addition, machines having large flywheels such as large punch presses.

- motors having high code letters and full voltage start; some older motors without code letters may be included in this group.

According to *Section 430-24* of the *NEC*, the minimum conductor ampacity required for the three appliances is 2.2 amperes + 3.96 amperes + (7.48 amperes $\times$ 1.25) = 15.51 amperes. Therefore, the conductors to be used for the installation must have a minimum rating of 20 amperes, such as No. 12 TW conductors. A NEMA 15-20R receptacle, figure 9-7, is provided at each appliance location. Each appliance has a four-wire rubber cord with a NEMA 15-20P plug. This cord connects the appliance to the three-phase power, provides grounding to the appliance as required by *Section 250-43*, and serves as a disconnect means as required by *Sections 422-20* and *422-22*.

Each of the three appliances is purchased as a complete unit including a motor controller and motor overload protection. After the electrician provides the proper receptacle outlets, the appliances are ready for use as soon as they are moved into place and plugged into the outlets.

Fig. 9-7 A NEMA 15-20R receptacle and plate.

THE DOUGHNUT MACHINE

An additional individual branch circuit provides power to the doughnut machine. The electrical equipment on the doughnut machine consists of (1) a 2,000-watt heating element which heats the liquid used in frying and (2) a driving motor which has a full-load rating of 2.2 amperes.

This doughnut machine is purchased as a complete, prewired unit equipped with a four-wire cord to be plugged into the proper receptacle. Since it may be necessary to check the equipment before it is installed, it is essential that the electrician be able to read and understand control circuit diagrams such as the one shown in figure 9-8. The following components are indicated on the control circuit for the doughnut machine (figure 9-8).

S   A manual switch used to start and stop the machine.

T1   A thermostat with its sensing element in the frying tank; this thermostat keeps the oil at the correct temperature.

T2   Another thermostat with its sensing element in the frying tank; this thermostat controls the drive motor.

A   A three-pole contactor controlling the heating element.

B   A three-pole motor controller operating the drive motor.

M   A three-phase motor.

OL   Overload units installed for overload protection.

P   Pilot light to indicate when power is *On*.

Although this appliance is on an individual circuit, its current draw is limited to 80% of the branch-circuit rating according to *NEC Section 384-16* (see Unit 4). The branch circuit supplying the doughnut machine must have sufficient ampacity to meet the minimum load requirements as calculated according to *NEC Section 210-23*.

Heater load	= 2,000 W
Motor load	
2.2 amperes $\times$ 208 volts	
$\times$ 1.73	= 792 VA
25% of 792 VA	= 198 VA
Total	2,990 VA

The maximum load permitted on a 20-ampere, three-phase branch circuit is 16 amperes $\times$ 208

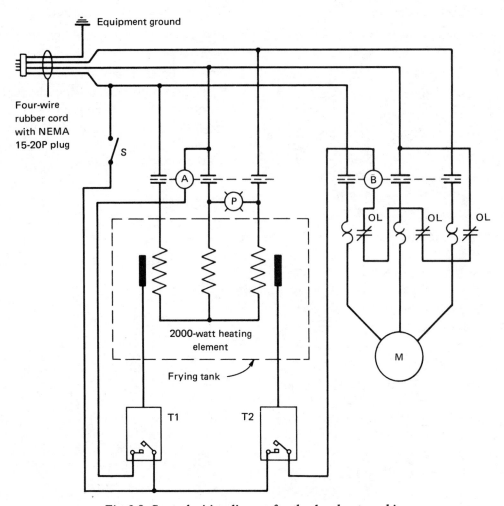

Fig. 9-8 Control wiring diagram for the doughnut machine.

volts × 1.73 = 5,760 volt-amperes. The load of 2,990 volt-amperes is well within the permitted loading of a 20-ampere circuit.

THE BAKE OVEN

The final appliance installed in the bakery is a bake oven with a rated load of 16,000 watts. The oven is classified as a permanently connected appliance located to be on a specific circuit. It consists of the oven proper, a step-down transformer, and a wall-mounted contactor controlled by a thermostat sensitive to the internal temperature of the oven, figure 9-9.

Disconnect Means

A fusible disconnect switch is installed as specified by the contract for the electrical work in the bakery, figures 9-9 and 9-10. This switch satisfies the requirements of *NEC Section 422-21* gov-

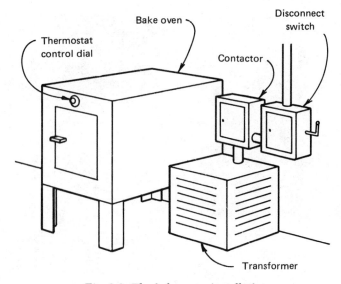

Fig. 9-9 The bake oven installation.

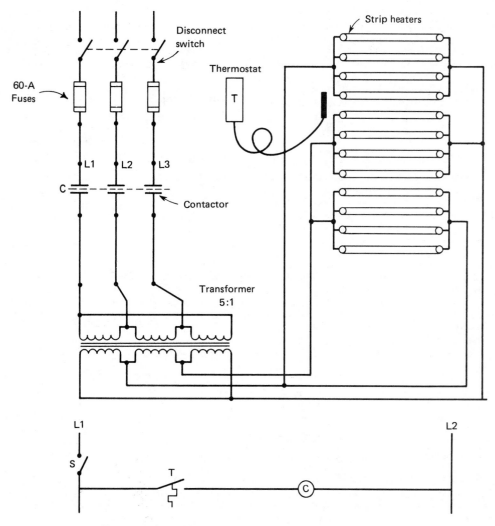

Fig. 9-10 Wiring and control diagram for the bake oven.

erning the disconnecting means for permanently connected appliances.

Overcurrent Protection

Article 450 of the *NEC* covers the correct installation procedures for a transformer, figure 9-11. The following calculations show how the size of the branch circuit for the installation is determined.

The full load rating is:

$$\text{Amperes} = \frac{\text{W}}{\text{Volts} \times 1.73} = \frac{16,000}{208 \times 1.73}$$
$$= 44.4 \text{ Amperes}$$

The maximum connected load permitted on a 50-ampere branch circuit is 40 amperes according to *Section 384-16* of the *NEC*. As a result, a 60-ampere circuit is required. However, the maximum

Fig. 9-11 Dry-type transformer.

permissible overcurrent protection for this type of installation according to *Section 450-3* is:

44.4 amperes × 1.25 = 55.5 amperes

Section 450-3(b)(1), Exception No. 1, permits the selection of the next higher standard fuse rating which, in this case, is 60 amperes. Therefore, 60-ampere fuses will be installed in the disconnect switch. (See Unit 2, figure 2-2.)

The electrician must be certain to use time-delay fuses when applying fuses on transformer installations sized at 125% of the transformer's rated current. Fast acting, current-limiting types of fuses could blow when attempting to energize the transformer. A discussion on the time-current characteristics of fuses and circuit breakers is covered in Unit 18.

Grounding

Fixed equipment can be grounded using the methods listed in *Section 250-91* of the *NEC*. If flexible metal conduit is used as the connecting raceway, then a conductor must be installed in the raceway so that the proper grounding is obtained (refer to figure 8-18).

——REVIEW——

Note: Refer to the *National Electrical Code* or the plans as necessary.

1. A furnace is a _____ appliance.

2. An electric 1/4-inch drill motor is a _____ appliance.

3. A 50-gallon water heater is a _____ appliance, and is

    _____ .

4. A commercial-type hair dryer is a _____ appliance, and is _____ .

5. An attachment plug and receptacle may serve as the disconnect means for an appliance. True _____ False _____

6. Circuit No. 4 is a general-purpose branch circuit because _____

    _____

    _____ .

7. The actual load on each conductor supplying the doughnut machine is _____ ampere(s).

8. When the bake oven is heating, the current in the secondary of the bake oven transformer is _____ amperes.

9. If an insulated conductor is used as the grounding means, it shall have a(an) _____ color.

10. The maximum size permitted for the fuse used as the running overload protection on a 50°C motor rated at 6 amperes is _____ ampere(s).

11. Although motors have many different starting characteristics that can vary the amount of inrush current they draw when started, it generally is possible to install time-delay, dual-element fuses sized at _____ % of a motor's normal, full-load running current.

12. *Section 450-3* of the *NEC* states that overcurrent protection for a transformer is not to exceed _____ % of the transformer's rated current.

UNIT 10

Reading Electrical Drawings—
Insurance Office

OBJECTIVES

After completing the study of this unit, the student will be able to

- list (tabulate) the material required to install an electrical wiring system.

PRINTS

A good electrician is able to look at an electrical drawing and prepare a list or tabulation of the material required to install the wiring system. A great deal of time can be lost if the proper materials are not on the job when needed. It is essential that the electrician prepare in advance for the installa-

tion so that the correct variety of material is available in sufficient quantities to complete the job.

LOADING SCHEDULE

The loading schedule for the insurance office is included at this point to assist the student in reviewing the project.

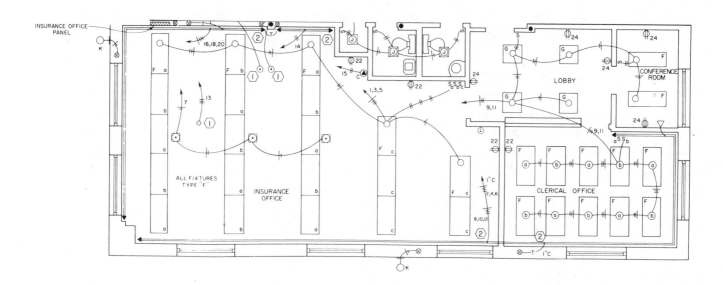

NOTE: FOR COMPLETE BLUEPRINT, REFER TO BLUEPRINT E-3.

INSURANCE OFFICE LOADING SCHEDULE

Loading	NEC Article	NEC Minimum (watts)	Actual Allowance (watts)	Design Value
GENERAL LIGHTING				
1,374 sq. ft. × 3.5 watts/sq. ft.	220-2(b)	4,809		
32 Style F luminaires @ 200 watts			6,400	
4 Style G luminaires @ 100 watts			400	
Total general lighting		4,809	6,800	
Value to be used				6,800 watts
MOTOR LOAD				
CM 20.2 amperes × 1.73 × 208 volts		7,272	7,272	
EFM 3.2 amperes × 208 volts		665	665	
CFM 3.2 amperes × 208 volts		665	665	
25% allowance of CM motor	440-7, 440-32	1,818		
Total motor load		10,420	8,602	
Value to be used				10,420 watts
OTHER LOAD				
Receptacle outlets				
15 outlets @ 180 volt-amperes/outlet	220-2(b)	2,700		
Copier outlet				
30 amperes × 0.8 × 208	384-16(c)	4,992		
Multioutlet assembly				
120 ft. × 1.5 amperes/ft. × 120 volts	220-2(c)	21,600		
Total other load		29,292		
Value to be used				29,292 watts
Total for area				46,512 watts

APPLICATION QUESTIONS

1. List the different types of lamps used in the insurance office and the number required.
   a. Type _____ number required _____
   b. Type _____ number required _____

2. Indicate the number of switches used in the insurance office.
   a. Single-pole switches _____
   b. Three-way switches _____
   c. Four-way switches _____

3. Indicate the number of receptacles used in the insurance office.
   a. NEMA 5-15R single _____
   b. NEMA 5-15R duplex _____
   c. NEMA 5-20R duplex _____
   d. NEMA 6-20R duplex _____
   e. NEMA 15-20R single _____
   f. NEMA 6-30R single _____

4. Branch circuit No. 15 serves the receptacle outlet for the copy machine.
   a. How many feet (meters) of conduit are required to install this circuit? _____

   _____

   b. How many box connectors are required in the installation? _____

5. Branch circuits No. 9 and No. 11 serve the lighting in the southeast area of the office. Determine the total amounts used of the following materials.
   a. Switch boxes _____
   b. 4-inch octagonal boxes _____
   c. Conduit _____ feet (_____ meters)
   d. Box connectors _____
   e. White wire _____ feet (_____ meters)
   *f. Black wire _____ feet (_____ meters)
   *g. Red wire _____ feet (_____ meters)
   *h. Blue wire _____ feet (_____ meters)
   i. 4-inch blank plates _____
   j. 3/8-inch flexible conduit _____ feet (_____ meters)
   k. 3/8-inch flexible conduit box connectors _____
   l. White fixture wire _____ feet (_____ meters)
   m. Black fixture wire _____ feet (_____ meters)

   *Although color coding is not required by the *National Electrical Code,* the use of colored conductors is highly recommended when it is necessary to keep track of specific circuits.

UNIT 11

The Cooling System

OBJECTIVES

After completing the study of this unit, the student will be able to

- list the parts of a cooling system.

- describe the function of each part of the cooling system.

- make the necessary calculations to obtain the sizes of the electrical components.

- read a typical wiring diagram which shows the operation of a cooling unit.

The electrician working on commercial construction is expected to install the wiring of cooling systems and troubleshoot electrical problems in these systems. Therefore, it is recommended that the electrician know the basic theory of refrigeration and the terms associated with it.

REFRIGERATION

Refrigeration is a method of removing energy in the form of heat from an object. When the heat is removed, the object is colder. An energy balance is maintained which means that the heat must go somewhere. As long as the locations where the heat is discharged and where it is absorbed are remote from each other, it can be said that the space where the heat was absorbed is cooled. The inside of the household refrigerator or freezer is cold to the touch, but this cold cannot be used to cool the kitchen by having the refrigerator door open. Actually, leaving the door open causes the kitchen to become hotter. This situation demonstrates an important principle of mechanical refrigeration: to remove heat energy, it is necessary to add energy or power to it.

Mechanical refrigeration relies primarily on the process of evaporation. This process is responsible for the cool sensation which results when rubbing alcohol is applied to the skin or when gasoline is spilled on the skin. Body heat supplies the energy required to vaporize the alcohol or gasoline. It is the removal of this energy from the body that causes the sensation of cold. In refrigeration systems such as those used in the commercial building, the evaporation process is controlled in a closed system. The purpose of this arrangement is to preserve the refrigerant so that it can be reused many times in what is known as the *refrigerant cycle*. As shown in figure 11-1, the four main components of the refrigerant cycle are:

- *Evaporator*
  The refrigerant evaporates here as it absorbs energy from the removal of heat.

- *Compressor*
  This device raises the energy level of the refrigerant so that it can be condensed readily to a liquid.

- *Condenser*
  The compressed refrigerant condenses here as the heat is removed.

- *Expansion valve*
  This metering device maintains a sufficient unbalance in the system so that there is a point of low pressure where the refrigerant can expand and evaporate.

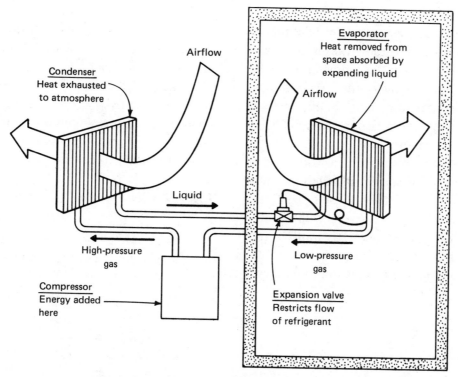

Fig. 11-1 The refrigerant cycle.

EVAPORATOR

The evaporator in a commercial installation normally consists of a fin tube coil, figure 11-2, through which the building air is circulated by a motor-driven fan. A typical evaporator unit is shown in figure 11-3. The evaporator may be located inside or outside the building. In either case, the function of the evaporator is to remove heat from the interior of the building or the enclosed space. The air is usually circulated through pipes or ductwork to insure a more even distribution. The window-type air conditioner, however, discharges the air directly from the evaporator coil. In general, the cooling air from the evaporator is recirculated within the space to be cooled and is passed again across the cooling coil. A certain percentage of outside air is added to the circulating air to replace air lost through exhaust systems and fume hoods, or because of the gradual leakage of air through walls, doors, and windows.

COMPRESSOR

The compressor serves as a pump to draw the expanded refrigerant gas from the evaporator. In addition, the compressor boosts the pressure of the

(A) Fin tube coil

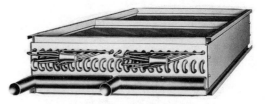

(B) Fin tube evaporator coil

Fig. 11-2 Evaporator components.

Fig. 11-3 An evaporator with dual fans.

gas and sends it to the condenser, figure 11-4. The compression of the gas is necessary since this process adds the heat necessary to cause the gas to be condensed to a liquid. When the temperature of the air or water surrounding the condenser is relatively warm, the gas temperature must be increased to insure that the temperature around the condenser will liquefy the refrigerant.

Direct-drive compressors are usually used in large installations. For smaller installations, however, the trend is toward the use of hermetically sealed compressors. Due to several built-in electrical characteristics, these hermetic units cannot be used on all installations. (Restrictions to the use of hermetically sealed compressors are covered later in this text.)

CONDENSER

Condensers are generally available in three types: as air-cooled units, figure 11-5, as water-cooled units, figure 11-6, or as evaporative cooling units, figure 11-7, in which water from a pump is sprayed on air-cooled coils to increase their capacity. The function of the condenser in the refrigerant cycle is to remove the heat taken from the evaporator, plus the heat of compression. Thus, it can be seen that keeping the refrigerator door open causes the kitchen to become hotter because the

condenser rejects the combined heat load to the condensing medium. In the case of the refrigerator in a residence, the condensing medium is the room air.

Air-cooled condensers use a motor-driven fan to drive air across the condensing coil. Water-cooled condensers require a pump to circulate the water. Once the refrigerant gas is condensed to a liquid state, it is ready to be used again as a coolant.

EXPANSION VALVE

It was stated previously that the refrigerant must evaporate or boil if it is to absorb heat. The process of boiling at ordinary temperatures can occur

Fig. 11-4 A motor-driven reciprocating compressor.

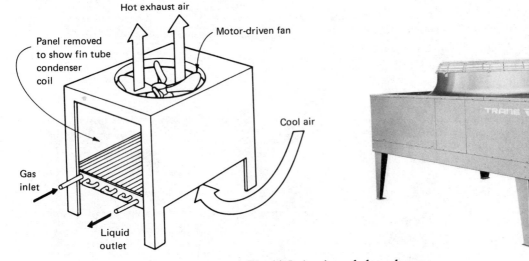

Fig. 11-5 An air-cooled condenser.

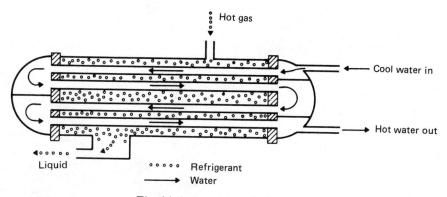

Fig. 11-6 **A water-cooled condenser.**

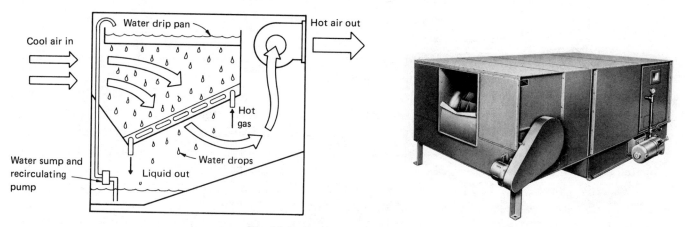

Fig. 11-7 **An evaporative condenser.**

in the evaporator only if the pressure is reduced. The task of reducing the pressure is simplified since the compressor draws gas away from the evaporator and tends to evacuate it. In addition, a restricted flow of liquid refrigerant is allowed to enter the high side of the evaporator. As a result, the pressure remains fairly low in the evaporator coil so that the liquid refrigerant entering through the restriction flashes into a vapor and fills the evaporator as a boiling refrigerant.

The restriction to the evaporator may be a simple orifice. In commercial systems, however, a form of automatic expansion valve is generally used because it is responsive to changes in the heat load. The expansion valve is located in the liquid refrigerant line at the inlet to the evaporator. The valve controls the rate at which the liquid flows into the evaporator. The liquid flow rate is determined by installing a temperature-sensitive bulb at the outlet of the evaporator to sense the heat gained by the refrigerant as it passes through the evap-

orator. The bulb is filled with a volatile liquid (which may be similar to the refrigerant). This liquid expands and passes through a capillary tube connected to a spring-loaded diaphragm to cause the expansion valve to meter more or less refrigerant to the coil. The delivery of various amounts of refrigerant compensates for changes in the heat load of the evaporator coil.

HERMETIC COMPRESSORS

The tremendous popularity of mechanical refrigeration for household use in the 1930s and 1940s stimulated the development of a new series of nonexplosive and nontoxic refrigerants. These refrigerants are known as chlorinated hydrofluorides of the halogen family and, at the time of their initial production, were relatively expensive. The expense of these refrigerants was great enough so that is was no longer possible to permit the normal leakage which occurred around the shaft seals of reciprocating, belt-driven units. As a result, the

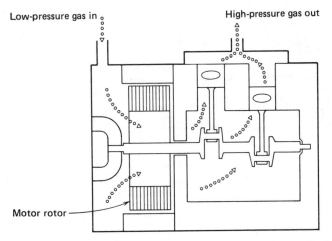

Low-pressure gas in

High-pressure gas out

Motor rotor

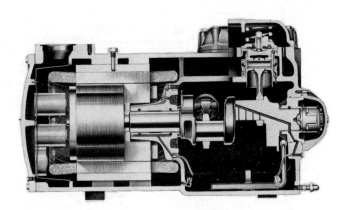

Fig. 11-8 The hermetic compressor.

hermetic compressor was developed, figure 11-8. This unit consists of a motor-compressor completely sealed in a gastight, steel casing. The refigerant gas is circulated through the compressor and over the motor windings, rotor, bearings, and shaft. The circulation of the expanded gas through the motor helps to cool the motor.

The initial demand for hermetic compressors was for use on residential-type refrigerators. Therefore, most of the hermetic compressors were constructed for single-phase service. In other words, it was necessary to provide auxiliary winding and starting devices for the compressor installation. Because the refrigerant gas surrounded and filled the motor cavity, it was necessary to remove the centrifugal switch commonly provided to disconnect the starting winding at approximately 85 percent of the full speed. The switch was removed because any arcing in the presence of the refrigerant gas caused the formation of an acid from the hydrocarbons in the gas. This acid attacked and etched the finished surfaces of the shafts, bearings, and valves. The acid also carbonized the organic material used to insulate the motor winding and caused the eventual breakdown of the insulation. To overcome the problem of the switch, the relatively heavy magnetic winding of a relay was connected in series with the main motor winding. The initial heavy inrush of current caused the relay to lift a magnetic core and energize the starting winding. As the motor speed increased, the main winding current decreased and allowed the relay to remove the starting or auxiliary winding from the circuit.

A later refinement of this arrangement was the use of a voltage-sensitive relay which was wound to permit pickup at values of voltage greater than the line voltage. The coil of this voltage-sensitive relay was connected across the starting winding. This connection scheme was based on the principle that the rotor of a single-phase induction motor induces in its own starting winding a voltage which is approximately in quadrature (phase) with the main winding voltage and has a value greater than the main voltage. The voltage-sensitive relay broke the circuit to the starting winding and remained in a sealed position until the main winding was de-energized.

Since the major maintenance problems in hermetic compressor systems were due generally to starting relays and capacitors, it was desirable to eliminate as many as possible of these devices. It was soon realized that small- and medium-sized systems could make use of a different form of refrigerant metering device, thus eliminating the automatic expansion valve. Recall that this valve has a positive shutoff characteristic which means that the refrigerant is restricted when the operation cycle is finished as indicated by the evaporator reaching the design temperature. As a result, when a new refrigeration cycle begins, the compressor must start against a head of pressure. If a small bore, open capillary tube is substituted for the expansion valve, the refrigerant is still metered to the evaporator coil, but the gas continues to flow after the compressor stops until the system pressure is equalized. Therefore, the motor is only required to start the compressor against the same pressure on each side of the pistons. This ability to decrease the load led to the development of a new series of motors that contained only a running capacitor (no relay was installed). This type of motor furnished sufficient torque to start the unloaded compressor and, at the same time, greatly improved the overall power factor.

COOLING SYSTEM CONTROL

Figure 11-9 shows a wiring diagram that is representative of standard cooling units.

When the selector switch in the heating-cooling control is set to COOL and the fan switch is placed on AUTO, any rise in temperature above the set point causes the cooling contacts (TC) to close and complete two circuits. One circuit through the fan switch energizes the evaporator fan relay (EFR) which, in turn, closes contacts EFR to complete the 208-volt circuit to the evaporator fan motor (EFM). The second circuit is to the control relay (CR) and causes the CR1 contacts to open. When the CR1 contacts open, the crankcase heater (CH) is de-energized. (This heater is installed to keep the compressor oil warm and dry when the unit is not running.) In addition, CR2 contacts close to complete a circuit to the condenser fan motor (CFM) and another circuit through low-pressure switch 2 (LP2), the high-pressure switch (HP), the thermal contacts (T), and the overload contacts (OL) to the compressor motor starter coil (CS). When the CR2 contacts close, the three CS contacts in the power circuit to the compressor motor (CM) also close.

Contacts LP1 and LP2 open when the refrigerant pressure drops below a set point. When contacts LP2 open, CS is de-energized. When contacts LP1 open, CR is de-energized, the circuit to CFM opens, and the circuit to CH is completed. The high-pressure switch (HP) contacts open and de-energize the compressor motor starter (CS) when the refrigerant pressure is above the set point. The low-pressure control (LP1) is the normal operating control and the high-pressure control (HP) and LP2 act as safety devices. The T contacts shown in figure 11-9 are located in the compressor motor and open when the winding temperature of the motor is too

high. The OL contacts are controlled by the OL elements installed in the power leads to the motor. The OL elements are sized to correspond to the current draw of the motor. That is, a high current causes the overload elements to overheat and open the OL contacts. As a result, the compressor starter is de-energized and the compressor stops. The evaporator fan motor (EFM) used to circulate air in the store area can be run continuously if the fan switch is turned to FAN. Actually, in many situations, it is recommended that the fan motor run continuously to keep the air in motion.

COOLING SYSTEM INSTALLATION

The owner of the commercial building leases the various office and shop areas on the condition that heat will be furnished to each area. However, tenants agree to pay the cost of operating the refrigeration system to provide cooling in their areas. Only four cooling systems are indicated in the plans for the commercial building since the bakery does not use a cooling system. The cooling equipment for the insurance office, the doctor's office, and the beauty salon are single-package cooling units located on the roof. The compressor, condenser, and evaporator for each of these units are constructed within a single enclosure. The system for the drugstore is a split system with the compressor and condenser located on the roof and the evaporator located in the basement, figure 11-10. Regardless of the type of cooling system installed, a duct system must be provided to connect the evaporator with the area to be cooled. The duct system shown diagrammatically in figure 11-10 does not represent the actual duct system which will be installed by another contractor.

The electrician is expected to provide a power circuit to each air-conditioning unit as shown on the plans. In addition, it is necessary to provide wiring to the thermostat in each area, and, in the case of the drugstore, wiring must be provided to the evaporator located in the basement.

DISCONNECT SWITCH

The requirements of the air-conditioning power circuits are covered in detail in the units dealing with branch circuits. These circuits will

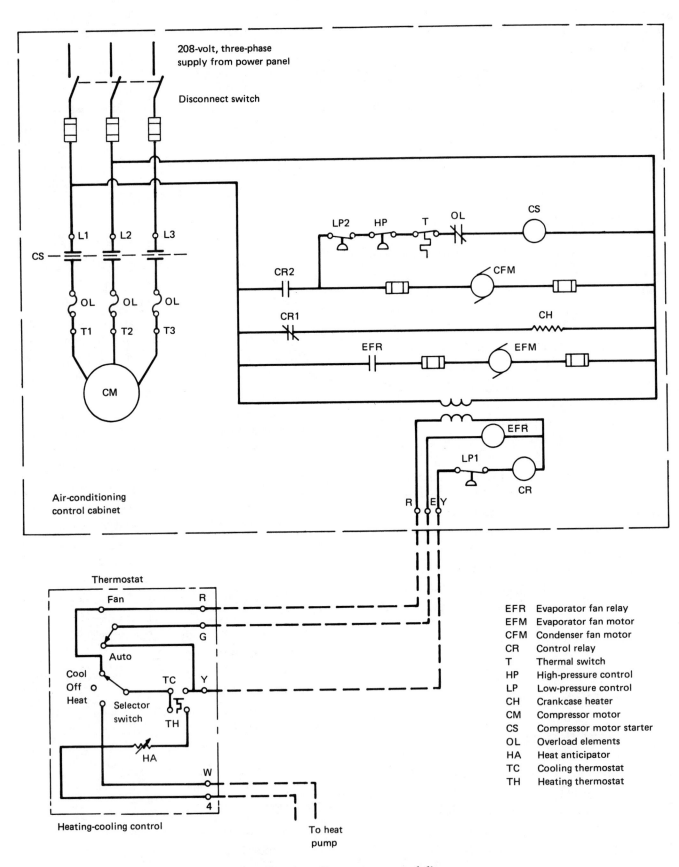

Fig. 11-9 A cooling system control diagram.

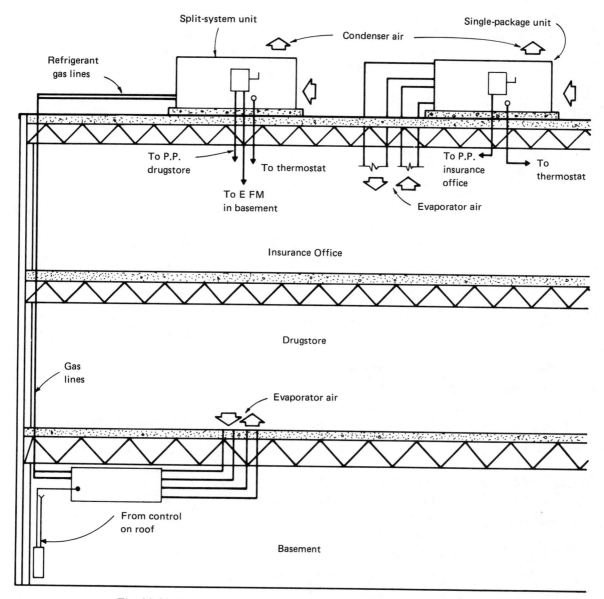

Fig. 11-10 Single-package cooling units and a split-system cooling unit.

terminate in a raintight, nonfused disconnect switch installed adjacent to the cooling units located on the roof. The disconnect switch must meet the requirements of *Section 430-102.* The selection of the disconnect switch is governed by *Section 440-12.* This section specifies that a disconnect switch must be:

- rated for at least 15 percent of the nameplate load current(s) or the branch-circuit selection current, whichever is greater.

- rated greater than the locked rotor current(s) as determined from *NEC Table 430-151.*

- located within sight of the motor.

- readily accessible.

The following example shows how *Section 440-12* is applied to the insurance office air-conditioning unit to find the full-load current.

Compressor motor (CM): full-load current (FLA), 20.2 amperes
locked rotor current (LRA), 90 amperes
Evaporator fan motor (EFM): 1/4 hp, 3.2 amperes, 208 volts single phase
Condenser fan motor (CFM): 1/4 hp, 3.2 amperes, 208 volts single phase
Unit total FLA: 26.6 amperes

According to the *Notes* for *NEC Tables 430-148* and *430-150,* the 230-volt full-load current values given in the tables must be increased by 10%

Table 430-148.   Full-Load Currents in Amperes

Single-Phase Alternating-Current Motors

The following values of full-load currents are for motors running at usual speeds and motors with normal torque characteristics. Motors built for especially low speeds or high torques may have higher full-load current, and multispeed motors will have full-load current varying with speed, in which case the nameplate current ratings shall be used.

To obtain full-load currents of 208- and 200-volt motors, increase corresponding 230-volt motor full-load currents by 10 and 15 percent, respectively.

The voltages listed are rated motor voltages. The currents listed shall be permitted for system voltage ranges of 110 to 120 and 220 to 240.

HP	115V	230V
1/6	4.4	2.2
1/4	5.8	2.9
1/3	7.2	3.6
1/2	9.8	4.9
3/4	13.8	6.9
1	16	8
1½	20	10
2	24	12
3	34	17
5	56	28
7½	80	40
10	100	50

Table 430-150.   Full-Load Current*
Three-Phase Alternating-Current Motors

	Induction Type Squirrel-Cage and Wound-Rotor Amperes					Synchronous Type †Unity Power Factor Amperes			
HP	115V	230V	460V	575V	2300V	230V	460V	575V	2300V
½	4	2	1	.8					
¾	5.6	2.8	1.4	1.1					
1	7.2	3.6	1.8	1.4					
1½	10.4	5.2	2.6	2.1					
2	13.6	6.8	3.4	2.7					
3		9.6	4.8	3.9					
5		15.2	7.6	6.1					
7½		22	11	9					
10		28	14	11					
15		42	21	17					
20		54	27	22					
25		68	34	27		53	26	21	
30		80	40	32		63	32	26	
40		104	52	41		83	41	33	
50		130	65	52		104	52	42	
60		154	77	62	16	123	61	49	12
75		192	96	77	20	155	78	62	15
100		248	124	99	26	202	101	81	20
125		312	156	125	31	253	126	101	25
150		360	180	144	37	302	151	121	30
200		480	240	192	49	400	201	161	40

For full-load currents of 208- and 200-volt motors, increase the corresponding 230-volt motor full-load current by 10 and 15 percent, respectively.

* These values of full-load current are for motors running at speeds usual for belted motors and motors with normal torque characteristics. Motors built for especially low speeds or high torques may require more running current, and multispeed motors will have full-load current varying with speed, in which case the nameplate current rating shall be used.

† For 90 and 80 percent power factor the above figures shall be multiplied by 1.1 and 1.25 respectively.

The voltages listed are rated motor voltages. The currents listed shall be permitted for system voltage ranges of 110 to 120, 220 to 240, 440 to 480, and 550 to 600 volts.

Table 430-151.   Conversion Table of Locked-Rotor Currents for Selection of Disconnecting Means and Controllers as Determined from Horsepower and Voltage Rating

For use only with Sections 430-110, 440-12 and 440-41.

Motor Locked-Rotor Current Amperes*							Max. HP Rating
Single Phase			Two or Three Phase				
115V	230V	115V	200V	230V	460V	575V	
58.8	29.4	24	18.8	12	6	4.8	½
82.8	41.4	33.6	19.3	16.8	8.4	6.6	¾
96	48	43.2	24.8	21.6	10.8	8.4	1
120	60	62	35.9	31.2	15.6	12.6	1½
144	72	81	46.9	40.8	20.4	16.2	2
204	102	—	66	58	26.8	23.4	3
336	168	—	105	91	45.6	36.6	5
480	240	—	152	132	66	54	7½
600	300	—	193	168	84	66	10
—	—	—	290	252	126	102	15
—	—	—	373	324	162	132	20
—	—	—	469	408	204	162	25
—	—	—	552	480	240	192	30
—	—	—	718	624	312	246	40
—	—	—	897	780	390	312	50
—	—	—	1063	924	462	372	60
—	—	—	1325	1152	576	462	75
—	—	—	1711	1488	744	594	100
—	—	—	2153	1872	936	750	125
—	—	—	2484	2160	1080	864	150
—	—	—	3312	2880	1440	1152	200

* These values of motor locked-rotor current are approximately six times the full-load current values given in Tables 430-148 and 430-150.

to obtain the full-load current values of 208-volt motors.

Thus, for the 1/4-hp, 208-volt, single-phase fan motors, the full-load current is:

2.9 × 1.1 = 3.19 amperes or 3.2 amperes

(Note that the full-load current of the air-conditioning unit is actually a *vector sum* (see Unit 4).

To determine the disconnect switch rating, refer to *Table 430-150*. In the 230-volt column, find a current value which is equal to or greater than the total air-conditioning unit full-load current of 26.6 amperes. Since this value is 28 amperes, a 10-hp disconnect switch is required for the air conditioner.

The total equivalent locked rotor current is determined as follows:

Compressor motor: 90 LRA from manufacturer

Evaporator fan motor: 1/4 hp; using the lowest value in *Table 430-151*, the LRA is 29.4 amperes

Condenser fan motor: 1/4 hp; using the lowest value in *Table 430-151*, the LRA is 29.4 amperes

The total LRA is 148.8 amperes which requires a 10-hp disconnect switch according to *NEC Table 430-151* (see three-phase, 230-volt column). If one of the selection methods (FLA or LRA) indicates a larger size disconnect switch than the other method, then the larger switch should be installed.

COMPRESSOR MOTOR BRANCH-CIRCUIT PROTECTION

The rated full-load current for the compressor motor is given as 20.2 amperes. Since this motor is a hermetic motor, and separate running overcurrent protection is provided, the rating of the motor branch-circuit protection is determined from *Section 440-22(a)* of the *NEC*. This section requires that the motor-compressor branch-circuit protective device:

- be capable of carrying the starting current.

- have a rating of not greater than 175% of the rated load current or of the branch-circuit selection current (whichever is greater).

- be rated at 225% of the rated load current if the 175% rating does not permit starting.

- be rated at 15 amperes minimum.

Since the rated full-load current for the compressor motor is 20.2 amperes, 20.2 amperes $\times$ 1.75 = 35.35 amperes. Since this value is greater than 35 amperes, the next higher standard size of protective device may be used, as long as the value of the device is less than 225% of the full-load current. Therefore, for the compressor motor, a 40-ampere protective device is permissible. However, if a dual-element fuse is to be used, then the 35-ampere value is preferred since it permits starting and gives added protection.

The motor running overcurrent protective device is selected according to the requirements of *Section 440-52* of the *NEC*. This section requires that the motor-compressor be protected from overloads and failure to start by:

- a separate overload relay which is responsive to current flow and trips at not more than 140% of the rated load current, or

- an integral thermal protector arranged so that its action interrupts the current flow to the motor-compressor, or

- a time-delay fuse or inverse time circuit breaker rated at not more than 125% at the rated load current.

For the equipment in the commercial building, a separate overload relay rated at not more than 28.28 amperes is selected (20.2 amperes $\times$ 1.40 = 28.28 amperes).

Overload relays generally are available in a variety of sizes from the manufacturer of the motor starter (controller). Each manufacturer usually lists in the starter cover the recommended devices to be used for various full-load ampere ratings. Caution should be exercised in selecting a device from this list since it is normally based on 115% to 125% of the full-load rating. These values will not provide sufficient protection for a hermetic motor. Manufacturers of motor starters often provide information on how to determine the actual trip currents. A quick calculation usually will indicate the proper sizing of the overload relay.

EVAPORATOR FAN MOTOR AND CONDENSER FAN MOTOR RUNNING OVERLOAD PROTECTION

The motors for the evaporator fan and the condenser fan are the same size. As a result, the protective devices for these motors will also be the same. The branch-circuit overcurrent protective fuse will serve as the running overcurrent device. Since the motors are not compressors, the rating of the protective device will be determined by *Section 430-32(c)(1)*. If it is assumed that the fan motors have a service factor of 1.35 and a FLA of 3.2 amperes, then

3.2 amperes $\times$ 125% = 4.0 amperes

NEC Section 430-34 permits the use of the next higher fuse size if the 125% sizing does not permit the motor to start or to carry the load. The maximum size permitted is 140% for motors marked with a service factor of not less than 1.15.

3.2 $\times$ 1.40 = 4.48 amperes

(maximum size permitted)

However, a four-ampere dual-element time-delay fuse does have sufficient time delay for the starting and running characteristics of the evaporator fan motor (EFM) and for the condenser fan motor (CFM), and provides proper running overcurrent protection. See Table 9-1 in Unit 9 of this text for load response of fuses.

The nameplates of many types of electrical appliances supply *branch-circuit selection current ratings*. These ratings can be used to determine the branch-circuit ampacity. This method is simpler than adding the individual heaters, motors, and other loads within the appliances to obtain the branch-circuit rating.

—REVIEW—

Note: Refer to the *National Electrical Code* or the plans as necessary.

1. Refrigeration is a method of _____ from an object.

2. Match the following items by inserting the appropriate letter from Column II in the blank in Column I.

Column I		Column II
a. This device raises the energy level of the refrigerant.	_____	(A) Evaporator
b. The refrigerant condenses here as heat is removed.	_____	(B) Compressor
c. This is a metering device.	_____	(C) Condenser
d. The refrigerant absorbs energy here in the form of heat.	_____	(D) Expansion valve

3. Name three types of condensers. 1) _____ , 2) _____ , and 3) _____ .

4. A 230-volt single-phase hermetic compressor with an FLA of 21 and an LRA of 250 requires a disconnect switch with a rating of ____ hp.

5. A motor-compressor unit with a rating of 32 amperes is protected from overloads by a separate overload relay selected to trip at not more than ____ ampere(s).

UNIT 12

Special Systems

OBJECTIVES

After completing the study of this unit, the student will be able to

- install surface metal raceway.
- install multioutlet assemblies.
- install communication circuits.
- install floor outlets.

A number of electrical systems are found in almost every commercial building. Although these systems usually are a minor part of the total electrical work to be done, they are essential systems and it is recommended that the electrician be familiar with the installation requirements of these special systems.

SURFACE METAL RACEWAYS
(NEC Article 352)

Surface metal raceways are generally installed as extensions to an existing electrical raceway system, and where it is impossible to conceal conduits, such as in desks, counters, cabinets, and modular partitions. The installation of surface metal raceways is governed by *NEC Sections 352-1* through *352-8*. The number and size of the conductors to be installed in metal surface raceways is limited by the design of the raceway. Catalog data from the raceway manufacturer will specify the permitted number and size of the conductors for specific raceways. Conductors to be installed in raceways may be spliced at junction boxes or within the raceway if the cover of the raceway is removable. See *NEC Section 352-7*.

Surface metal raceways are available in various sizes, figure 12-1. A wide variety of special fittings makes it possible to use raceway in almost any situation. Figure 12-2 illustrates coupling, clip, and strap devices used to support surface raceway. Two examples of the use of surface raceway are shown in figures 12-3 and 12-4.

MULTIOUTLET ASSEMBLIES

The definition of a multioutlet assembly is given in *Article 100* of the *NEC*. In addition, the *NEC* requirements for multioutlet assemblies are

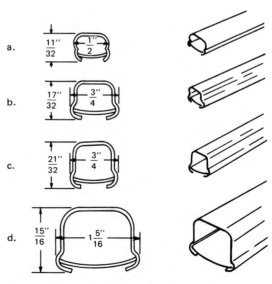

Fig. 12-1 Surface metal raceway.

Fig. 12-2 Raceway support.

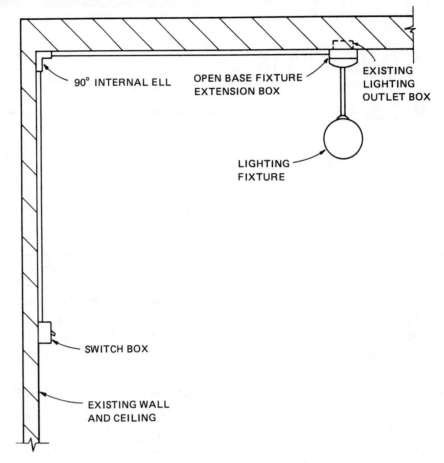

Fig. 12-3 Use of surface metal raceway to install switch on existing lighting installation.

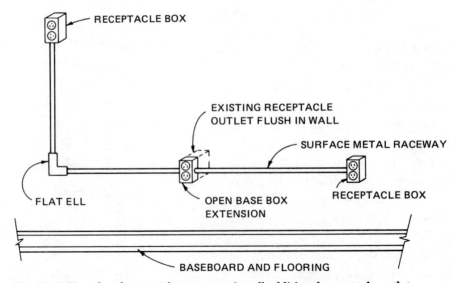

Fig. 12-4 Use of surface metal raceway to install additional receptacle outlets.

specified in *Article 353*. Multioutlet assemblies, figure 12-5, are similar to surface raceways and are designed to hold both conductors and devices. These assemblies offer a high degree of flexibility to an installation and are particularly suited to heavy use areas where many outlets are required or where there is a likelihood of changes in the installation requirements. The plans for the insurance office specify the use of a multioutlet assembly that will accommodate both power and communication cables, figure 12-6. This installation will allow the tenant in the insurance office to revise and expand the office facilities as the need arises.

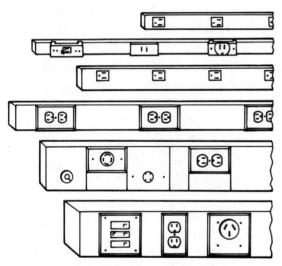

Fig. 12-5 Multioutlet assemblies.

Loading Allowance

The load allowance for a multioutlet assembly is specified by *NEC Section 220-2(c), Exception No. 1* as:

- 1 1/2 amperes for each 5 feet (1.52 m) of assembly when normal loading conditions exist, or

- 1 1/2 amperes for each 1 foot (305 mm) of assembly when heavy loading conditions exist.

The allowance for the insurance office is based on 120 feet (36.6 m) of multioutlet assembly at 1 1/2 amperes per foot:

120 ft. (36.6 m) × 1 1/2 amperes per foot
× 120 volts = 21,600 volt-amperes

The insurance office is provided with four 20-ampere circuits and six 15-ampere circuits which have a total capacity of only 16,320 volt-amperes. As a result, the maximum allowance of 21,600 volt-amperes means that there is room for expansion when necessary.

4 circuits × 1,920 volt-amperes per circuit	= 7,680 volt-amperes
6 circuits × 1,440 volt-amperes per circuit	= 8,640 volt-amperes
Total	16,320 volt-amperes

The No. 12 conductors in the 1-inch conduit feeding the multioutlet assembly are derated to 70% to comply with *Note 8* to *Table 310-16* of the *NEC*. This derating is made as it is expected that additional conductors will be added to the conduit. The derating of a 20-ampere conductor to 70% of

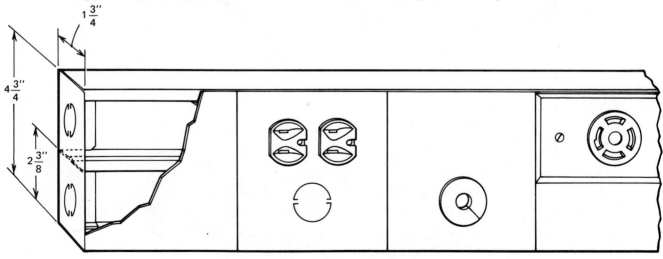

Fig. 12-6 Multioutlet assembly for power and communication systems.

its rating means that a 15-ampere protective device must be installed. This device can be loaded only to 80% of its capacity. As a result, the maximum permissible load per circuit is:

$$80\% \times 15 \text{ amperes} \times 120 \text{ volts} =$$
$$1{,}440 \text{ volt-amperes}$$

Receptacle Wiring

The plans indicate that the receptacles to be mounted in the multioutlet assembly must be spaced 30 inches (762 mm) apart. The receptacles may be connected in either of the arrangements shown in figure 12-7. That is, all of the receptacles on a phase can be connected in a continuous row, or the receptacles can be connected on alternate phases.

COMMUNICATION SYSTEMS

The installation of the telephone system in the commercial building will consist of two separate installations. The *electrical contractor* will install an empty conduit system according to the specifi-

cations for the commercial building and in the locations indicated on the plans. In addition, the installation will meet the rules, regulations, and requirements of the telephone company that will serve the building. Once the conduit system is complete, the telephone company will install a complete telephone system.

The telephone company installation includes pulling cables into the empty conduits and connecting all equipment necessary to provide a telephone (communication) system as required by the needs of the tenants and owner of the building.

Power Requirements

An allowance of 2,500 volt-amperes is made for the installation of the telephone equipment to provide the power required to operate the special switching equipment for a large number of telephones. Because of the importance of the telephone as a means of communication, the receptacle outlets for this equipment are connected to the emergency power system.

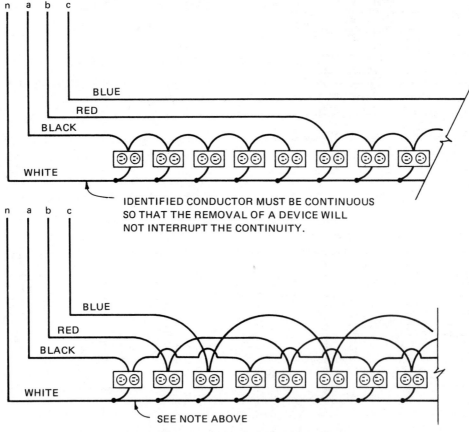

Fig. 12-7 Connection of receptacles.

The Telephone System

Telephone service requirements vary widely according to the type of business and the extent of the communication convenience desired. In many situations, the telephone lines may be installed in exposed locations. However, improvements in building construction techniques have made it more important to provide facilities for installing concealed telephone wiring. The use of new wall materials and wall insulation, the reduction or the complete omission of trim around windows and doors, the omission of baseboards, and the increased use of metal trim make it more difficult to attach exposed wires. For these reasons (and many others), the wiring is more conspicuous if the installation is exposed. In addition, the unprotected wiring is more subject to subsequent faulty operation.

To solve the problems of exposed wiring, conduits are installed and the proper outlet boxes and junction boxes are placed in position during the construction. The material and construction costs are low when compared to the benefits gained from this type of installation. The telephone company maintains a service known as the *Architects and Builders Service* to supply the necessary information to owners, architects, and builders to enable them to plan the installation of the raceways (conduits) in the proper locations to conceal the telephone installation. The telephone company provides the services of its engineers without cost.

Since it is generally more difficult to conceal the telephone wiring to the floors above the first floor after the building construction is completed, it is recommended that telephone conduits be provided to these floors. The materials used in the installation and the method of installing the conduits for the telephone lines are the same as those used for light and power wiring with the following modifications.

- Because of small openings and limited space, junction boxes are used rather than the standard conduit fittings such as ells and tees.

- Since multipair conductors are used extensively in telephone installations, the size of the conduit should be 3/4 inch or larger. The current and potential ratings for an installation of this type are not governed by the same conditions as conventional wiring for light and power.

- The number of bends and offsets is kept to a minimum; when possible, bends and offsets are made using greater minimum radii or larger sweeps (the allowable minimum radius is 6 inches, or 152 mm).

- A fishwire is installed in each conduit for use by the telephone company personnel in pulling in the cables.

- When basement telephone wiring is to be exposed, the conduits are dropped into the basement and terminated with a bushing (junction boxes are not required). No more than 2 inches (50.8 mm) of conduit should project beyond the joists or ceiling level in the basement.

- The inside conduit drops need *not* be grounded unless they may become energized (*Section 250-80*).

- The service conduit carrying telephone cables from the exterior of a building to the interior must be permanently and effectively grounded, *Sections 250-80* and *800-11(c)(4)*.

Installing the Telephone Outlets

The plans for the commercial building indicate that each of the occupancies requires telephone service. As shown previously, the insurance office uses a multioutlet assembly. Telephone outlets are installed in the balance of the occupancies. These outlets are supplied by EMT which runs to the basement, figure 12-8, where the cable then runs exposed to the main terminal connections. Telephone company personnel will install the cable and connect the equipment.

FLOOR OUTLETS

In an area the size of the insurance office, it may be necessary to place equipment and desks where wall outlets are not available. Floor outlets may be installed to provide the necessary electrical supply to such equipment. Two methods can be used to provide floor outlets: (1) installing underfloor raceway or (2) installing floor boxes.

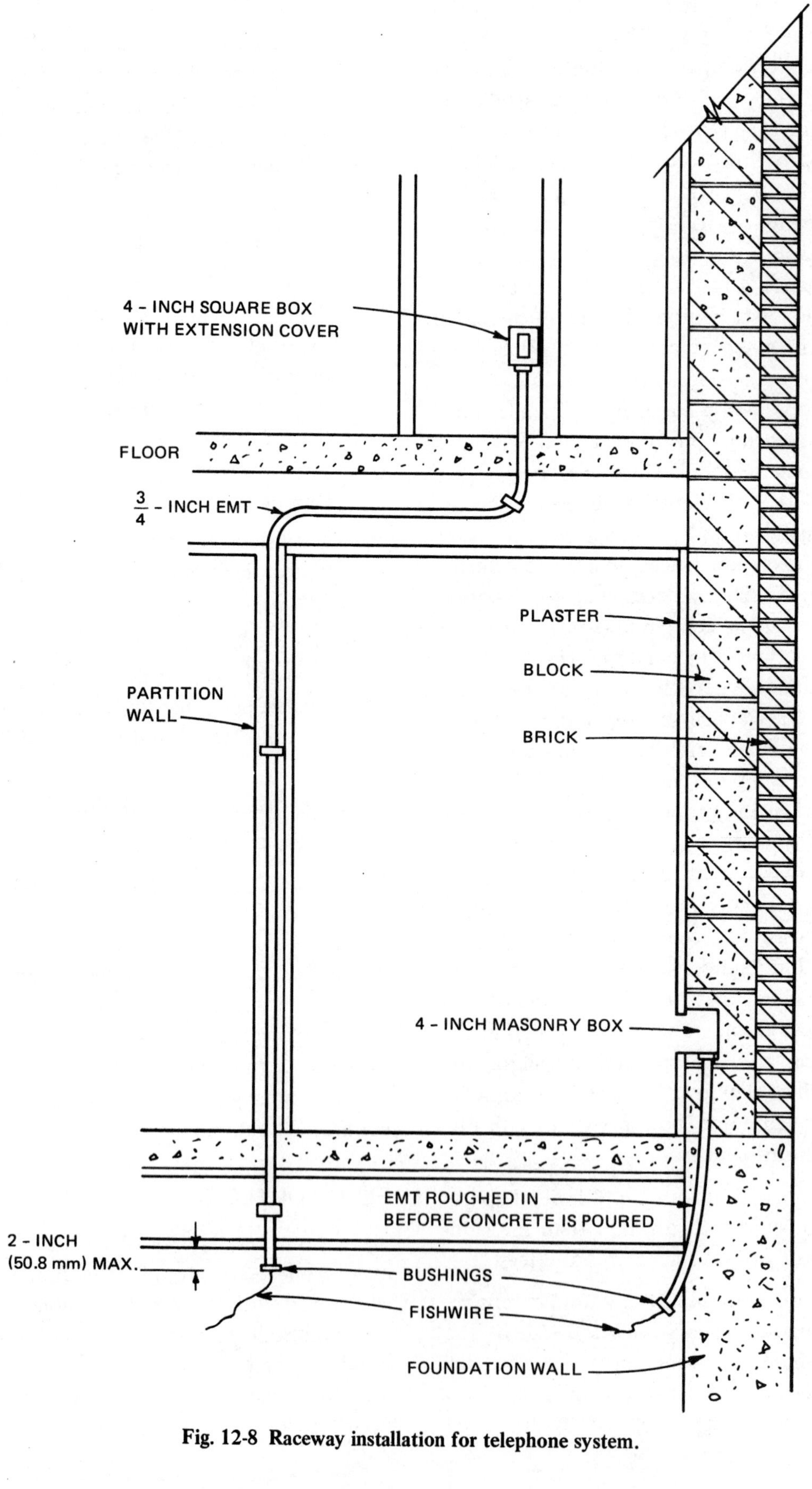

Fig. 12-8 Raceway installation for telephone system.

Underfloor Raceway (*NEC Article 354*)

In general, underfloor raceway is installed to provide both power and communication outlets using a dual duct system similar to the one shown in figure 12-9. The junction box is constructed so that the power and communication systems are always separated from each other. Service fittings are available for the outlets, figure 12-10.

Floor Boxes

Floor boxes can be installed using any approved raceway such as rigid conduit or EMT. Before the concrete floor is poured, the box must be installed to the correct height by adjusting its leveling screws, figure 12-11.

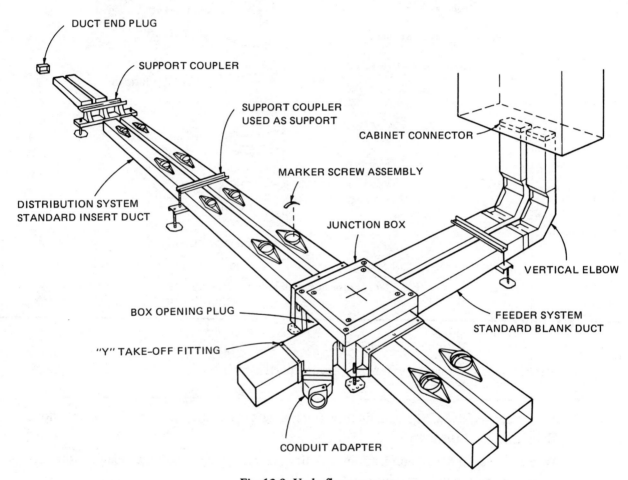

Fig. 12-9 Underfloor raceway.

FOR HIGH POTENTIAL SERVICE

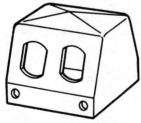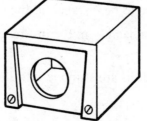

Dimensions: $4\frac{1}{8}''$ (105 mm) long; $4\frac{1}{8}''$ (105 mm) wide; $2\frac{15}{16}''$ (74.6 mm) high.

Fig. 12-10 Service fittings.

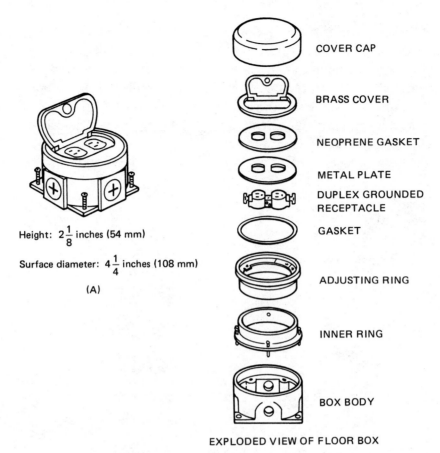

Height: $2\frac{1}{8}$ inches (54 mm)

Surface diameter: $4\frac{1}{4}$ inches (108 mm)

(A)

COVER CAP

BRASS COVER

NEOPRENE GASKET

METAL PLATE

DUPLEX GROUNDED RECEPTACLE

GASKET

ADJUSTING RING

INNER RING

BOX BODY

EXPLODED VIEW OF FLOOR BOX

(B)

Fig. 12-11 Floor box with leveling screws.

REVIEW

Note: Refer to the *National Electrical Code* or the plans as necessary.

1. Surface raceway may be extended through partitions if the _____
   _____ .

2. In surface raceway, the conductors (including taps and splices) shall fill not more than ____ % of the area.

3. Where power and communications circuits are run in combination raceway, the different systems shall be run in _____ .

4. In areas of light usage, a _____ -ampere capacity is allowed for each _____ feet (_____ meters) of multioutlet assembly.

5. How many floor outlets are to be installed in the insurance office? _____

6. To install the multioutlet assembly in the insurance office, _____ end caps and _____ internal ells are required.

7. How many telephone conduits are stubbed into the basement? _____

UNIT 13

Reading Electrical Drawings— Beauty Salon

OBJECTIVES

After completing the study of this unit, the student will be able to

- make branch-circuit calculations.
- lay out the raceway system.
- tabulate the material requirements for the installation.

ELECTRIC WATER HEATER

A beauty salon uses a large amount of hot water. To provide this hot water, the specifications indicate that the circuits to the beauty salon are to supply an electric water heater. The water heater is not furnished by the electrical contractor but is to be connected by the contractor as indicated in the specifications, figure 13-1. The water heater is rated at 4,000 watts and 208 volts, and will be permanently connected to a separate circuit.

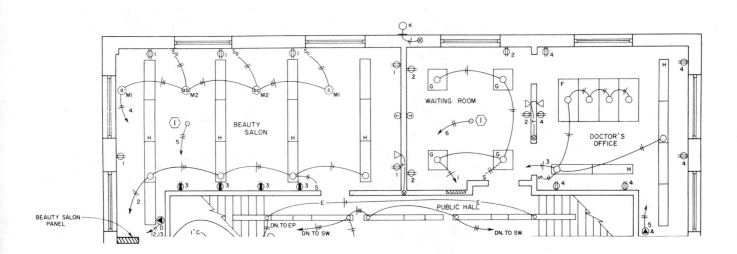

NOTE: FOR COMPLETE BLUEPRINT, REFER TO BLUEPRINT E-3.

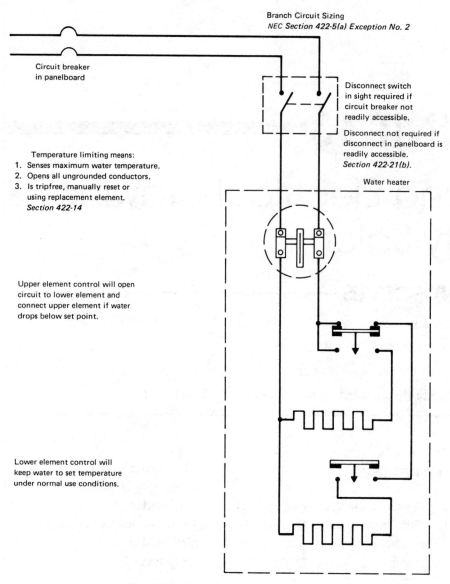

Branch Circuit Sizing
NEC Section 422-5(a) Exception No. 2

Circuit breaker
in panelboard

Disconnect switch
in sight required if
circuit breaker not
readily accessible.

Disconnect not required if
disconnect in panelboard is
readily accessible.
Section 422-21(b).

Water heater

Temperature limiting means:
1. Senses maximum water temperature.
2. Opens all ungrounded conductors.
3. Is tripfree, manually reset or
   using replacement element.
   Section 422-14

Upper element control will open
circuit to lower element and
connect upper element if water
drops below set point.

Lower element control will
keep water to set temperature
under normal use conditions.

Fig. 13-1 Water heater installation.

APPLICATION QUESTIONS

A. Complete the following loading schedule by referring to the *National Electrical Code,* the plans, and the specifications.

BEAUTY SALON LOADING SCHEDULE			
Loading	*NEC* Minimum	Actual Allowance	Design Value
GENERAL LIGHTING			
_____ sq. ft. X _____ watts/sq. ft.			
_____ Style H luminaires @ 100 VA			
_____ Style M1 luminaires @ 150 VA			
_____ Style M2 luminaires @ 300 VA			
Totals			
Value to be used			
MOTOR LOAD			
Compressor			
_____ amperes X _____ X 208 volts			
Condenser-evaporator			
_____ amperes X _____ volts			
Allowance			
_____ % of larger motor			
Totals			
Value to be used			
OTHER LOAD			
Water Heater			
_____ receptacle outlets @ _____ VA			
_____ receptacle outlets @ _____ VA			
Totals			
Value to be used			
Area Total			

B. 1. The water heater (can) (cannot) be connected to the electric power by a receptacle outlet. (Underline the correct answer.)

2. A _____-ampere circuit breaker or fuse should be installed for the water heater circuit.

3. A separate disconnect switch (is) (is not) required for the water heater. (Underline the correct answer.)

4. Make a list of the material required to install the raceway system for the beauty salon lighting circuits. Using Worksheet A, indicate in the blanks provided the quantities and any special requirements such as sizes and styles. Assume that type THHN wire will be installed.

WORKSHEET A

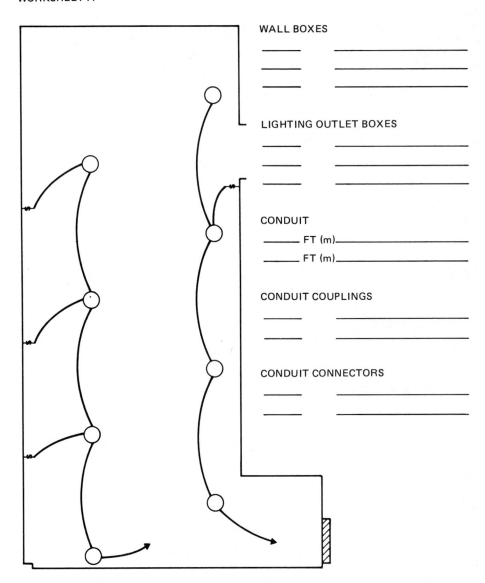

WALL BOXES

_____   _____
_____   _____
_____   _____

LIGHTING OUTLET BOXES

_____   _____
_____   _____
_____   _____

CONDUIT
_____ FT (m)_____
_____ FT (m)_____

CONDUIT COUPLINGS

_____   _____
_____   _____

CONDUIT CONNECTORS

_____   _____
_____   _____

5. Using Worksheet B, draw the configurations for the two styles of receptacles used in the beauty salon. It (is) (is not) necessary to use the two styles of receptacles. (Underline the correct answer.)

WORKSHEET B

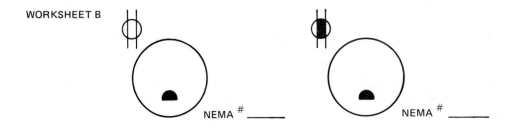

NEMA #_____   NEMA #_____

UNIT 14

Lamps for Lighting

OBJECTIVES

After completing the study of this unit, the student will be able to

- list the types of lamps used in the commercial building.
- define the technical terms relating to lamp selection and installation.
- list the parts of three types of lamps.
- list the operating characteristics of lamp types.
- recognize the significance of lamp designations.

For most construction projects, the electrician is required to purchase and install lamps in the luminaires (lighting fixtures). Three types of lamps are specified for the commercial building: (1) incandescent, (2) fluorescent, and (3) mercury. Thomas Edison provided the talent and perseverance that led to the development of the incandescent lamp in 1879 and the fluorescent lamp in 1896. Peter Cooper Hewitt produced the mercury lamp in 1901. All three of these lamp types have been refined and greatly improved since they were first developed.

Code *Article 410* contains the provisions for the wiring, and installation of lighting fixtures, lampholders, lamps, receptacles, and rosettes.

LIGHTING TERMINOLOGY

Candela (cd): The luminous intensity of a source, when expressed in candelas, is the candlepower (cp) rating of the source.

Lumen (lm): The amount of light received in a unit of time on a unit area at a unit distance from a point source of one candela, figure 14-1. The surface area of a sphere with unit radius is 12.57 times the radius;

therefore, a one-candela source produces 12.57 lumens. When the measurement is in English units, the unit area is 1 square foot and the unit distance is 1 foot. If the units are in SI, then the unit area is 1 square meter and the unit distance is 1 meter.

Illuminance: The measure of illuminance on a surface is the lumen per unit area expressed in

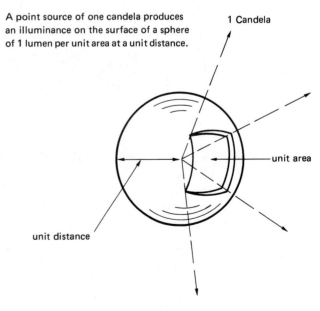

A point source of one candela produces an illuminance on the surface of a sphere of 1 lumen per unit area at a unit distance.

1 Candela

unit area

unit distance

Fig. 14-1

footcandles (fc) or lux (lx). The recommended illuminance levels vary greatly, depending on the task to be performed and the ambient lighting conditions. For example, while 5 footcandles (fc), or 54 lux (lx), is often accepted as adequate illumination for a dance hall, 200 fc (2,152 lx) may be necessary on a drafting table for detailed work.

Lumen per watt (lm/W): The measure of the effectiveness (efficacy) of a light source in producing light from electrical energy. A 100-watt incandescent lamp producing 1,670 lumens has an effective value of 16.7 lumens per watt.

Kelvin (K): The kelvin (sometimes incorrectly called degree Kelvin) is measured from absolute zero; it is equivalent to a degree value in the Celsius scale plus 273.16.

INCANDESCENT LAMPS

The incandescent lamp has the lowest efficacy of the three types of lamps listed previously. However, incandescent lamps are very popular and account for more than 50 percent of the lamps sold in the United States. This popularity is due largely to the low cost of incandescent lamps and luminaires.

Construction

The light-producing element in the incandescent lamp is a tungsten wire called the *filament*,

figure 14-2. This filament is supported in a glass envelope or bulb. The air is evacuated from the bulb and is replaced with an inert gas such as argon. The filament is connected to the base by the lead-in wires. The base of the incandescent lamp supports the lamp and provides the connection means to the power source. The lamp base may be any one of the base styles shown in figure 14-3.

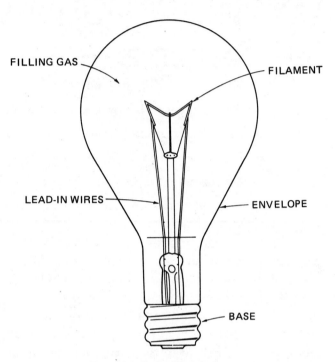

Fig. 14-2 Incandescent lamp.

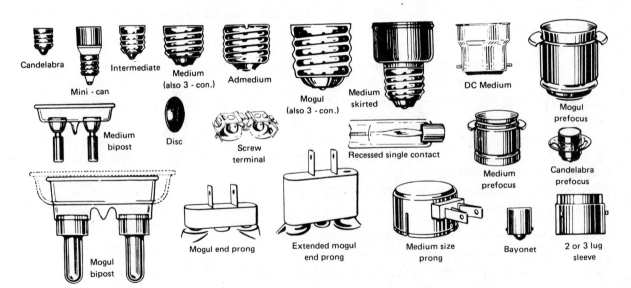

Fig. 14-3 Incandescent lamp bases.

Characteristics

Incandescent lamps are classified according to the following characteristics.

Voltage Rating. Incandescent lamps are available with many different voltage ratings. One manufacturer lists 33 different voltage ratings ranging from 5.5 to 500 volts in its large lamp catalog. When installing lamps, the electrician should be sure that a lamp with the correct rating is selected since a small difference between the rating and the actual voltage has a great effect on the lamp life and lumen output, figure 14-4.

Wattage. Lamps are usually selected according to their wattage rating. This rating is an indication of the consumption of electrical energy but is not a true measure of light output. For example, at the rated voltage, a 60-watt lamp produces 840 lumens and a 300-watt lamp produces 6,000 lumens; therefore, one 300-watt lamp produces more light than seven 60-watt lamps.

Shapes. Figure 14-5 illustrates the common lamp configurations and their letter designations.

Size. Lamp size is usually indicated in eighths of an inch and is the diameter of the lamp at the widest place. Thus, the lamp designation A19 means that the lamp has an arbitrary shape and is 19/8 inches or 2 3/8 inches (60.3 mm) in diameter, figure 14-6.

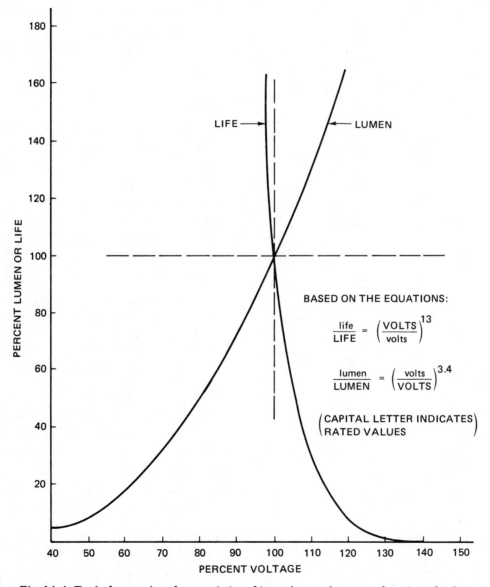

BASED ON THE EQUATIONS:

$$\frac{life}{LIFE} = \left(\frac{VOLTS}{volts}\right)^{13}$$

$$\frac{lumen}{LUMEN} = \left(\frac{volts}{VOLTS}\right)^{3.4}$$

$$\left(\begin{array}{l}\text{CAPITAL LETTER INDICATES}\\ \text{RATED VALUES}\end{array}\right)$$

Fig. 14-4 Typical operating characteristics of incandescent lamp as a function of voltage.

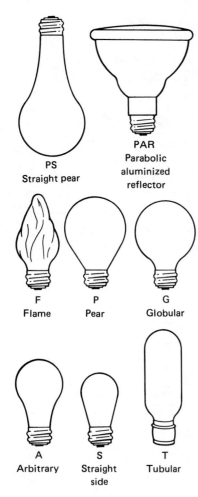

Fig. 14-5 Common lamp shapes.

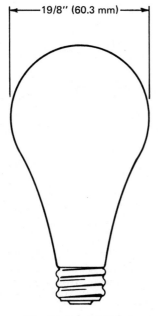

Fig. 14-6 An A19 lamp.

Operation

The light-producing filament in an incandescent lamp is a resistance load that is heated to a high temperature by the electric current through the lamp. This filament is usually made of tungsten which has a melting point of 3,655 K. At this temperature, the tungsten filament produces light with an efficacy of 53 lumens per watt. However, to increase the life of the lamp, the operating temperature is lowered which also means a lower efficacy. For example, if a 500-watt lamp filament is heated to a temperature of 3,000 K, the resulting efficacy is 21 lumens per watt.

FLUORESCENT LAMPS (*NEC Article 410-Q*)

The *National Electrical Code* classifies luminaires using fluorescent lamps as electric-discharge lighting. Fluorescent lighting has the advantages of a high efficacy and long life.

Construction

A fluorescent lamp consists of a glass bulb with an electrode and a base at each end, figure 14-7. The inside of the bulb is coated with a phosphor (a fluorescing material), the air is evacuated, and an inert gas plus a small quantity of mercury is released into the bulb. The base styles for fluorescent lamps are shown in figure 14-8.

Characteristics

Fluorescent lamps are classified according to type, length or wattage, shape, and color.

Type. The lamps may be preheat, rapid start or instant start depending upon the ballast circuit.

Length or Wattage. Depending on the lamp type, either the length or the wattage is designated. For example, both the F40 preheat and the F48 instant start are 40-watt lamps, 48 inches (1.22 m) long. The bases of these two lamps are different, however. If built under SI specifications, these lamps would probably be 1,200 millimeters long.

Shapes. The fluorescent lamp usually has a straight tubular shape. Two exceptions are the circline lamp which forms a complete circle and the U-shaped lamp which is an F40T12 lamp having a

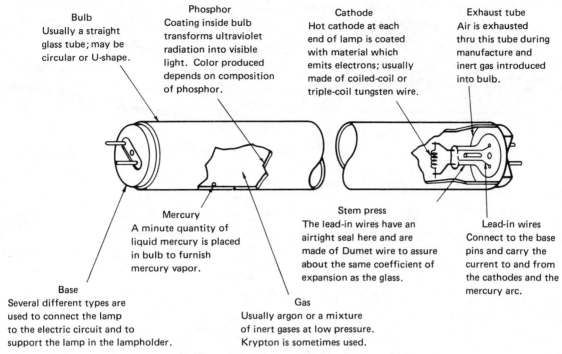

Bulb
Usually a straight glass tube; may be circular or U-shape.

Phosphor
Coating inside bulb transforms ultraviolet radiation into visible light. Color produced depends on composition of phosphor.

Cathode
Hot cathode at each end of lamp is coated with material which emits electrons; usually made of coiled-coil or triple-coil tungsten wire.

Exhaust tube
Air is exhausted thru this tube during manufacture and inert gas introduced into bulb.

Mercury
A minute quantity of liquid mercury is placed in bulb to furnish mercury vapor.

Stem press
The lead-in wires have an airtight seal here and are made of Dumet wire to assure about the same coefficient of expansion as the glass.

Lead-in wires
Connect to the base pins and carry the current to and from the cathodes and the mercury arc.

Base
Several different types are used to connect the lamp to the electric circuit and to support the lamp in the lampholder.

Gas
Usually argon or a mixture of inert gases at low pressure. Krypton is sometimes used.

Fig 14-7 Basic parts of a typical hot cathode fluorescent lamp.

180° bend in the center to fit a 2-foot (610 mm) long luminaire. A 2-foot lamp constructed under SI specifications would probably be 600 millimeters long.

Color. The color of a fluorescent lamp depends upon the phosphor mixture used to coat the inside of the lamp. The color is indicated in the lamp designation. For example, an F40T12CW is a cool white lamp and accents blue colors wherever it is used. Additional color designations are Warm White (WW), Daylight (D), White (W), Cool White Deluxe (CWX), and Warm White Deluxe (WWX). Decorative colors such as Blue (B), Pink (PK), and Green (G) are also available.

Operation

If a substance is exposed to such rays as ultraviolet and X rays and emits light as a result, then the substance is said to be fluorescing. The inside of the fluorescent lamp is coated with a phosphor material which serves as the light-emitting substance. When sufficient voltage is applied to the lamp electrodes, electrons are released. Some of these electrons travel between the electrodes to establish an electric discharge or arc through the mercury vapor in the lamp. As the electrons strike the

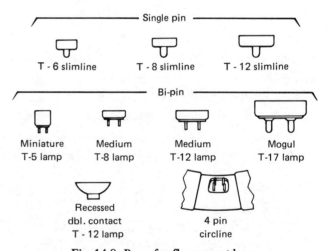

Fig. 14-8 Bases for fluorescent lamps.

mercury atoms, radiation is emitted by the atoms. This radiation is converted into visible light by the phosphor coating on the tube, figure 14-9.

As the mercury atoms are ionized, the resistance of the gas is lowered. The resulting increase in current ionizes more atoms. If allowed to continue, this process will cause the lamp to destroy itself. As a result, the arc current must be limited. The standard method of limiting the arc current is to connect a reactance (ballast) in series with the lamp.

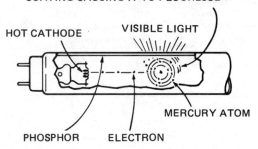

ULTRAVIOLET RADIATION STRIKES PHOSPHOR
COATING CAUSING IT TO FLUORESCE

HOT CATHODE

VISIBLE LIGHT

MERCURY ATOM

PHOSPHOR   ELECTRON

Fig. 14-9   How light is produced in a typical hot cathode fluorescent lamp.

Ballasts (*NEC Article 410-Q*)

Although inductive, capacitive, or resistive means can be used to ballast fluorescent lamps, the most practical ballast is an assembly of a core and coil, a capacitor, and a thermal protector installed in a metal case, figure 14-10. Once the assembled parts are placed in the case, it is filled with a potting compound to improve the heat dissipation and reduce ballast noise. Ballasts are available for the three basic operating circuits which are discussed next.

Preheat Circuit. The first fluorescent lamps developed were of the preheat type and required a starter in the circuit. This type of lamp is now obsolete and is seldom found except in smaller sizes which may be used for items such as desk lamps. The starter serves as a switch and closes the circuit until the cathodes are hot enough. The starter then opens the circuit and the lamp lights. The cathode temperature is maintained by the heat of the arc after the starter opens. Note in figure 14-11 that the ballast is in series with the lamp and acts as a choke to limit the current through the lamp.

Rapid-start Circuit. In the rapid-start circuit, the cathodes are heated continuously by a separate winding in the ballast, figure 14-12, with the result that almost instantaneous starting is possible. However, this type of fluorescent lamp requires the installation of a continuous grounded metal strip within an inch of the lamp. The metal wiring channel or the reflector of the luminaire can serve as this grounded strip. The standard rapid-start circuit operates with a lamp current of 430 mA.

Fig. 14-10   Fluorescent ballast.

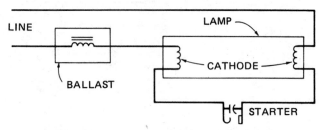

LINE   LAMP

BALLAST   CATHODE

STARTER

Fig. 14-11   Simple basic preheat circuit.

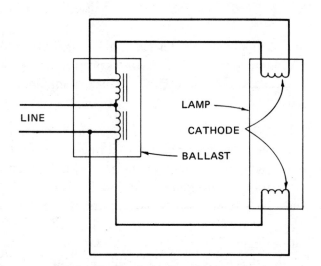

LINE   LAMP
                    CATHODE
                    BALLAST

Fig. 14-12   Simple basic rapid-start circuit.

However, two variations of the basic circuit are available. The high-output (HO) circuit operates with a lamp current of 800 mA and the very high-output circuit has 1,500 mA of current. Although high-current lamps are not as efficient as the standard lamp, they do provide a greater concentration of light, thus reducing the required number of luminaires.

Instant-start Circuit. The lamp cathodes in the instant-start circuit are not preheated. Sufficient voltage is applied across the cathodes to create an instantaneous arc, figure 14-13. As in the preheat

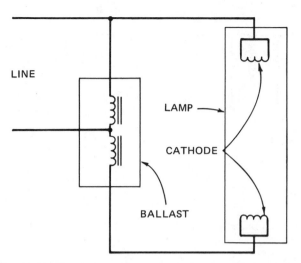

Fig. 14-13 Simple basic instant-start (cold cathode) circuit.

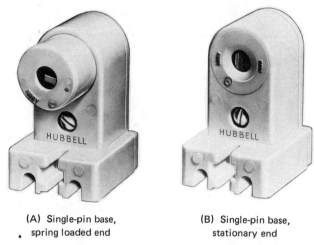

(A) Single-pin base, spring loaded end

(B) Single-pin base, stationary end

Fig. 14-14 Single-pin base for instant-start fluorescent lamp.

circuit, the cathodes are heated during lamp operation by the arc. The instant-start lamps require single-pin bases, figure 14-14, and are generally called *slimline lamps*. Bi-pin base fluorescent lamps are available, such as the 40-watt F40T12/CW/IS lamp. For this style of lamp, the pins are shorted together so that the lamp will not operate if it is mistakenly installed in a rapid-start circuit.

Special Circuits

Most fluorescent lamps are operated by one of the circuits just covered: the preheat, rapid-start, or instant-start circuits. Variations of these circuits, however, are available for special applications.

Dimming Circuit. The light output of a fluorescent lamp can be adjusted by maintaining a constant voltage on the cathodes and controlling the current passing through the lamp. Devices such as thyratrons, silicon-controlled rectifiers, and autotransformers can provide this type of control. The manufacturer of the ballast should be consulted about the installation instructions for dimming circuits. Dimming fluorescent lamps require a special dimming ballast.

Flashing Circuit. The burning life of a fluorescent lamp is greatly reduced if the lamp is turned on and off frequently. Special ballasts are available which maintain a constant voltage on the cathodes and interrupt the arc current to provide flashing.

High-frequency Circuit. Fluorescent lamps operate more efficiently at frequencies above 60 hertz. The gain in efficiency varies according to the lamp size and type. However, the gain in efficiency and the lower ballast cost generally are offset by the initial cost and maintenance of the equipment necessary to generate the higher frequency.

Direct-current Circuit. Fluorescent lamps can be operated on a dc power system if the proper ballasts are used. A ballast for this type of system contains a current-limiting resistor which provides an inductive kick to start the lamp.

Special Ballast Designation

Special ballasts are required for installations in cold areas such as out-of-doors. Generally, these ballasts are necessary for installations in temperatures lower than 10°C (50°F). These ballasts have a higher open-circuit voltage and are marked with the minimum temperature at which they will operate properly.

Class P. *[NEC Section 410-73(e)].* The National Fire Protection Association reports that the second most frequent cause of electrical fires in the United States is the overheating of fluorescent ballasts. To lessen this hazard, Underwriters Laboratories, Inc. has established a standard for a thermally protected ballast which is designated as a Class P ballast. This type of ballast has an internal protective device which is sensitive to the ballast temperature. This device opens the circuit to the ballast if the average

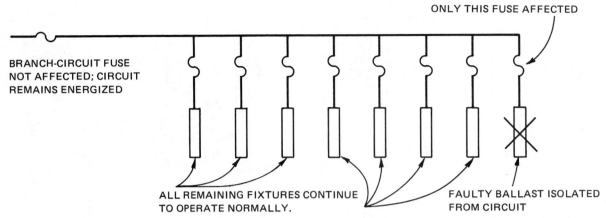

BRANCH-CIRCUIT FUSE
NOT AFFECTED; CIRCUIT
REMAINS ENERGIZED

ALL REMAINING FIXTURES CONTINUE
TO OPERATE NORMALLY.

FAULTY BALLAST ISOLATED
FROM CIRCUIT

Fig. 14-15 Each ballast is individually protected by fuse. In this system, a faulty ballast is isolated from the circuit.

ballast case temperature exceeds 90°C when operated in a 25°C ambient temperature. After the ballast cools, the protective device is automatically reset. As a result, a fluorescent lamp with a Class P ballast is subject to intermittent off-on cycling when the ballast overheats.

External fuses can be added for each ballast so that a faulty ballast can be isolated to prevent the shutdown of the entire circuit because of a single failure, figure 14-15. The ballast manufacturer normally provides information on the fuse type and its ampere rating. The specifications for the commercial building require that all ballasts shall be individually fused; the fuse size and type are selected according to the ballast manufacturer's recommendations, figure 14-16.

Section 410-73(e) requires that all fluorescent ballasts installed indoors, both for new installations and replacements, must have thermal protection built into the ballast.

Sound Rating. All ballasts emit a hum which is caused by magnetic vibrations in the ballast core. Ballasts are given a sound rating (from A to F) to indicate the severity of the hum. The quietest ballast has a rating of A. The need for a quiet ballast is determined by the ambient noise level of the location where the ballast is to be installed. For example, the additional cost of the A ballast is justified when it is to be installed in a doctor's waiting room. In the bakery work area, however, a ballast with a C rating is acceptable; in a factory, the noise of an F ballast probably will not be noticed.

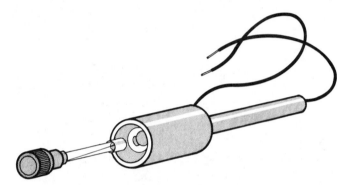

Fig. 14-16 In-line fuseholder for ballast protection.

Power Factor. The ballast limits the current through the lamp by providing a coil with a high reactance in series with the lamp. An inductive circuit of this type has an uncorrected power factor of from 40% to 60%; however, the power factor can be corrected to within 5% of unity by the addition of a capacitor. In any installation where there are to be a large number of ballasts, it is advisable to install ballasts with a high power factor.

Ballast Losses

A typical fluorescent ballast for two 40-watt lamps will draw a current of 0.8 amperes at 120 volts for a total of 96 volt-amperes. The difference between the lamp rating of 80 watts and the power input of 96 volt-amperes represents the ballast losses. These losses are both reactive and resistive and will vary with the ballast type and manufacturer. To assure a sufficient load capacity and to simplify the calculations, an allowance of 100 volt-amperes is usually made in load calculations for a luminaire using two 40-watt fluorescent lamps.

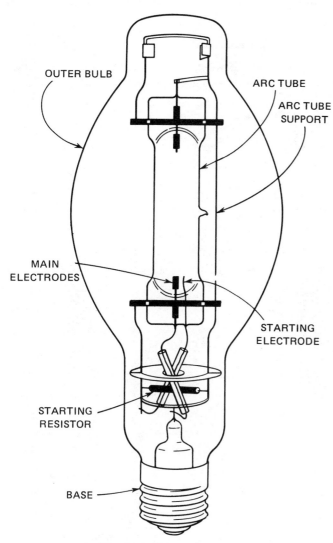

OUTER BULB

ARC TUBE

ARC TUBE SUPPORT

MAIN ELECTRODES

STARTING ELECTRODE

STARTING RESISTOR

BASE

Fig. 14-17 Mercury lamp.

MERCURY LAMPS

The mercury lamp is designated as a high-intensity electric discharge lamp (HID) and is noted for its good efficacy and long life. Because a mercury lamp has a long warm-up time before it reaches its full lumen output, it is generally considered unsuitable for interior office-type lighting.

Construction

The light from a mercury lamp is produced directly by the action of the arc on the vaporized mercury in the arc tube. The lamp consists of an inner glass envelope containing a quantity of mercury, argon gas, two main electrodes and a starting electrode. The inner glass tube is supported within an outer glass envelope or bulb. The outer bulb acts as a protective covering for the arc tube and helps to maintain a constant operating tem-

perature in the arc tube. The outer bulb may be phosphor-coated to improve color rendition. Mercury lamps usually are constructed with a base similar to the incandescent screw base, but in the larger mogul size.

Characteristics

Most mercury lamps are constructed using the BT (bulged-tubular) shape, figure 14-17. In recent years, improvements have been made in the color rendition of mercury lamps so that a good selection of colors is available. Mercury lamps are available in sizes from 50 to 3,000 watts and operate with an efficacy of from 30 lm/W to 60 lm/W.

When the lamp is turned on, a small arc is struck between the starting electrode and the main electrode next to it, figure 14-18. The small arc ionizes the argon gas to support an arc between the main electrodes. As the heat from the main arc begins to vaporize the mercury in the arc tube, it will take several minutes before the light output increases to its full brilliance. At this point, the mercury is completely vaporized. If a power interruption occurs while the mercury lamp is on, the lamp will not relight until the arc tube cools and the mercury condenses. A new arc cannot be struck by the available voltage until the vapor pressure decreases.

Lamp Designation

All mercury lamp designations begin with the letter H followed by numbers and letters to identify the ballast type, lamp size, and shape. To determine specific information about any mercury lamp designation, it is necessary to refer to lighting manuals or catalogs. For example, an H37-5KCIDX lamp is a 250-watt, E-28 (elongated 3 1/2-in, or 89-mm, diameter) deluxe white lamp.

Ballast

A mercury lamp also requires a ballast to provide the correct voltage and current for the lamp. The ballast can be mounted either within the luminaire or at a distance from the lamp. The mercury lamp ballasts shown in figure 14-19 are designed for remote mounting.

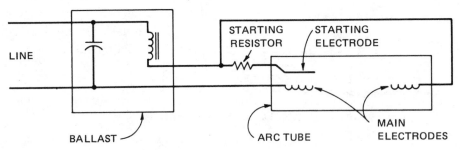

Fig. 14-18 Mercury lamp circuit.

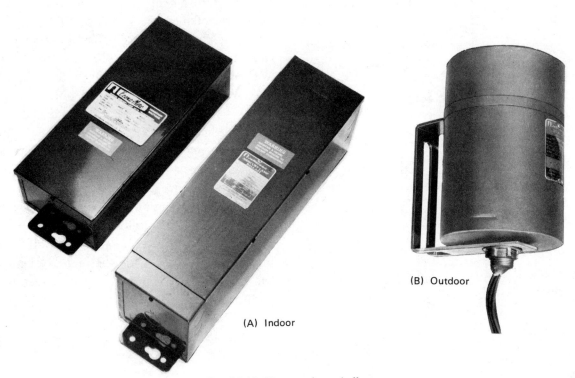

(A) Indoor

(B) Outdoor

Fig. 14-19 Mercury lamp ballasts.

METALLIC HALIDE LAMP

The metallic halide lamp is a variation of the mercury lamp. Although the metallic halide lamp has a higher efficacy (from 60 lm/W to 90 lm/W), it has a shorter average life and a high rate of light output depreciation over its lifetime.

SODIUM VAPOR LAMP

The low-pressure sodium vapor lamp has the highest efficacy (180 lm/W) of any light source presently available. However, the low-pressure sodium lamp has a predominantly yellow light which restricts its use to outdoor lighting. The high-pressure sodium vapor lamp has a better color rendition and an efficacy of 90-110 lm/W. The high-pressure lamp can be used in interior locations where color rendition is not critical.

XENON LAMP

Xenon lamps have good color rendition and instant start operation. However, they have a low efficacy of 25 lm/W to 50 lm/W and are available in very large wattages such as 15,000 W and above.

REVIEW

Note: Refer to the *National Electrical Code* or the plans as necessary.

1. Describe the lamps specified by the following lamp designations:
   a. 150 R _____
   b. 200 A _____
   c. 150 PAR _____
   d. F48T12/CW _____
   _____
   e. H37-5KC/DX _____

2. What is the size and shape of an A23 lamp? _____
   _____

3. The operating current of a 40-watt standard rapid-start lamp is _____ mA.

Match the following items.

4. Lumen per square foot _____   (A) 12.57 lumens
5. Lumen per watt _____   (B) Footcandle
6. One candela _____   (C) Efficacy
7. Starter _____   (D) Instant start
8. Cold cathode _____   (E) Rapid start
9. Constant heated cathode _____   (F) Preheat

10. The _____ is the light-producing element in the incandescent lamp.

11. The _____ is the light-producing element in the mercury lamp.

12. The _____ is the light-producing element in the fluorescent lamp.

13. The main purposes of a fluorescent ballast are to _____
    _____ .

14. When calculating the lighting load, _____ volt-amperes are allowed for a fluorescent luminaire with two 40-watt lamps.

15. Describe a Class P ballast. _____
    _____
    _____
    _____
    _____

16. Explain a method used to isolate a faulty ballast to prevent the shutdown of the entire circuit because of a single failure. _____
    _____
    _____
    _____
    _____

UNIT 15

Luminaires

OBJECTIVES

After completing the study of this unit, the student will be able to

- install a lighting outlet.
- install a lighting fixture.
- identify different types of luminaires and state their application.

DEFINITIONS

The terms *luminaire* and *lighting fixture* are used interchangeably. The Illuminating Engineering Society recommends the use of luminaire and the National Fire Protection Association in the *National Electrical Code* stresses the use of lighting fixture. The following definition applies to both terms.

A luminaire (lighting fixture) is a complete lighting assembly consisting of a lamp (or lamps) and components designed to distribute the light, position the lamp, and connect the lamp to the power supply.

A *lighting outlet* is defined as an outlet intended for the direct connection of a lampholder, a lighting fixture, or a pendant cord terminating in a lampholder.

INSTALLATION

The commercial electrician will find that the installation of lighting fixtures is a frequent part of work required for new building projects and remodeling projects where customers are upgrading the lighting in their properties. To complete work of this type, the electrician must know how to install fixtures and, in many cases, make the actual selection of the required lighting fixtures.

The luminaires required for the commercial building are described in the specification and indicated on the plans. These luminaires are divided into surface-mounting types and recessed-mounting types. The installation of lighting fixtures requires lighting outlets and supports. *NEC Sections 370-13, 410-1* and *410-10* through *410-16* govern the installation of the outlets and supports in what is known as the *rough-in*.

The rough-in is completed before the finish ceiling material is installed. The outlet boxes are located as shown on the plans. The exact location of the fixtures usually is not indicated on the plans. The electrician must locate the fixtures so that they are correctly spaced in the area. In special cases, the fixtures are located so that they illuminate one object or group of objects. If a single fixture is to be installed in an area, the center of the area can be found by drawing diagonals from each corner or by using the procedure shown in figure 15-1. When more than one fixture is located in an area, good light distribution is achieved by spacing the fixtures so that the distances between the fixtures and between the fixtures and the wall follow the recommended spacings given in figure 15-2. The spacing ratios for specific luminaires are given in the data sheets published by each manufacturer.

Supports

Both the lighting outlet and the lighting fixture must be supported from a structural member of the building. To provide this support, a large variety of clamps and clips is available, figures 15-3 and 15-4. The selection of the type of support depends upon the way in which the building is constructed.

Surface-mounted Fixtures. For surface-mounted fixtures, the lighting outlet must be roughed in at the proper location so that the fixture can be installed after the finish ceiling is applied. If a support is required for the fixture, then the previous statement applies to the support also. A support rod can be placed so that it extends about one inch (25.4 mm) below the finish ceiling. The support rod may be either a threaded rod or an unthreaded rod, figure 15-5. If actual fixtures are

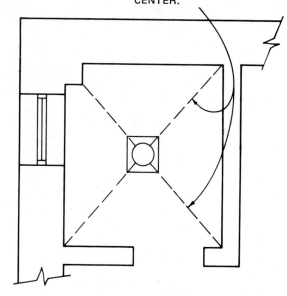

STRINGS STRETCHED FROM CORNERS TO LOCATE EXACT CENTER.

Fig. 15-1 Fixture location.

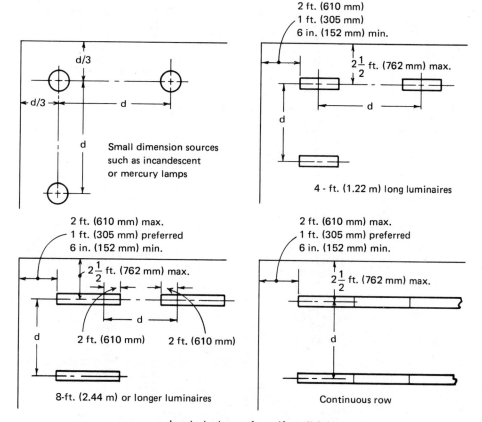

Luminaire layout for uniform lighting

Note: d should not exceed spacing ratio times mounting height. Mounting height is from luminaire to work plane for direct, semidirect, and general diffuse luminaires and from ceiling to work plane for semi - indirect and indirect luminaires.

Fig. 15-2 Fixture spacing.

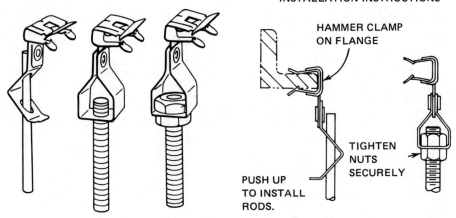

Fig. 15-3 Rod hangers for connection to flange.

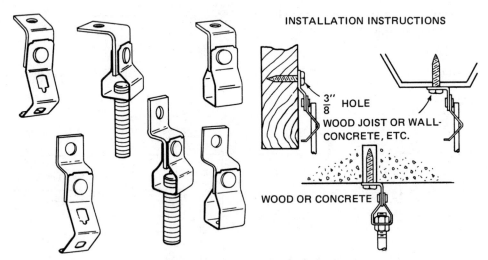

Fig. 15-4 Rod hanger supports for flat surfaces.

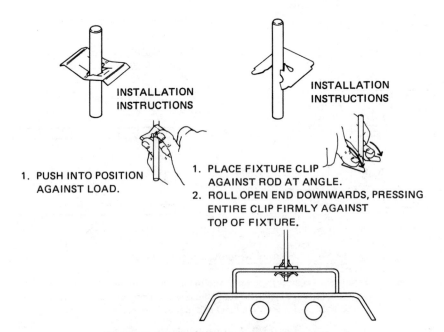

Fig. 15-5 Fixture support using unthreaded rod.

not available when the rough-in is done, it will be necessary to obtain the fixture drawings from the manufacturer. These drawings will indicate the exact dimensions of the mounting holes on the back of the fixture, figure 15-6.

Recessed Fixtures. In recessed fixtures, the light outlet box is located above the ceiling. The outlet box must be accessible to the fixture opening by means of a metal raceway that is at least four feet (1.22 m) long, but not more than six feet (1.83 m) long. Conductors suitable for the temperatures encountered are necessary. This information is marked on the fixture. Branch-circuit conductors may be run directly into the junction box on

approved prewired fixtures (*Section 410-67* and figure 15-7).

Recessed fixtures are usually supported by rails installed on two sides of the rough-in opening. Each rail can be heavy lather's channel or another type of substantial material.

Underwriters Laboratories provides the safety standards to which a fixture manufacturer must conform. Common fixture types are:

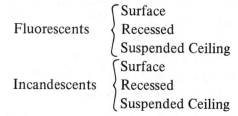

Fluorescents { Surface / Recessed / Suspended Ceiling

Incandescents { Surface / Recessed / Suspended Ceiling

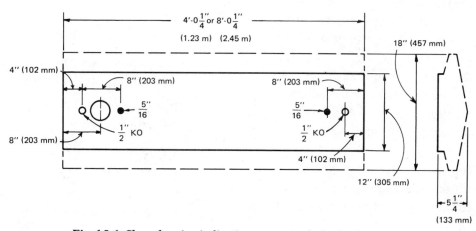

Fig. 15-6 Shop drawing indicating mounting holes for fixture.

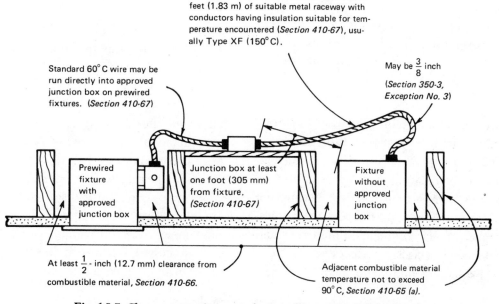

Fig. 15-7 Clearance requirements for installing recessed light fixtures.

IMPORTANT: Always carefully read the label on the fixture. The information found on the label, together with conformance to *Article 410* of the *NEC,* should result in a safe installation. The label provides the following information:

- wall mount only
- ceiling mount only
- maximum lamp wattage
- type of lamp
- access above ceiling required
- suitable for air handling use
- for chain or hook suspension only
- suitable for operation in ambients not exceeding _____°F (_____°C)
- suitable for installation in poured concrete
- for installation in poured concrete only
- for line volt-amperes, multiply lamp wattage by 1.25
- suitable for use in suspended ceilings
- suitable for use in noninsulated ceilings
- suitable for use in insulated ceilings
- suitable for damp locations (such as bathrooms and under eaves)
- suitable for wet locations
- suitable for use as a raceway
- suitable for mounting on low-density cellulose fiberboard

The Underwriters Laboratories Electrical Construction Materials Directory and the fixture manufacturers' catalogs and literature are excellent sources of information. The Underwriters Laboratories lists, tests and labels recessed fixtures as shown in figure 15-8.

Code Requirements for Installing Recessed Fixtures. The electrician must follow very carefully the Code requirements given in *Sections 410-64* through *410-72* for the installation and construction of recessed fixtures. Of particular importance are the Code restrictions on conductor temperature ratings, fixture clearances from combustible materials, and maximum lamp wattages. **Recessed fixtures generate a considerable amount of heat within the enclosure and are a definite fire hazard if not wired and installed properly,** figures 15-7 and 15-9.

A factor to be considered by the electrician when installing recessed fixtures is the necessity of working with the installer of the insulation to be sure that the clearances, as required by the Code, are maintained.

A boxlike device, figure 15-10, is available which snaps together around the recessed fixture, thus preventing the insulation from coming into contact with the fixture, as required by the Code. The material used in these boxes is fireproof.

According to *NEC Section 410-67(c),* a tap conductor must be run from the fixture terminal connection to an outlet box. For this installation, the following items must be true.

- The conductor insulation must be suitable for the temperatures encountered.

- The outlet box must be at least one foot (305 mm) from the fixture.

- The tap conductor must be installed in a suitable metal raceway.

- The raceway shall be at least four feet (1.22 m) but not more than six feet (1.83 m) long.

The branch-circuit conductors are run to this junction box where they are connected to conductors which have an insulation suitable for the temperature encountered at the fixture lampholder within the fixture. By locating the junction box at least one foot (305 mm) from the fixture, heat radiated directly from the fixture cannot overheat the wires in the junction box. Because the conductors rated for higher temperatures must run through at least four feet (1.22 m) of metal raceway (but not to exceed six feet, or 1.83 m) between the fixture and the junction box, any heat that may be conducted along the metal raceway will be dissipated considerably before reaching the junction box. Many recessed fixtures, especially the adjustable type, are factory equipped with a flexible metal raceway containing high temperature wires that meet the requirements of *Section 410-67(b).*

Recessed fixtures of the type installed in the shower and ceiling of the bathroom usually have a

TYPE	COMMENT
INSULATED CEILINGS 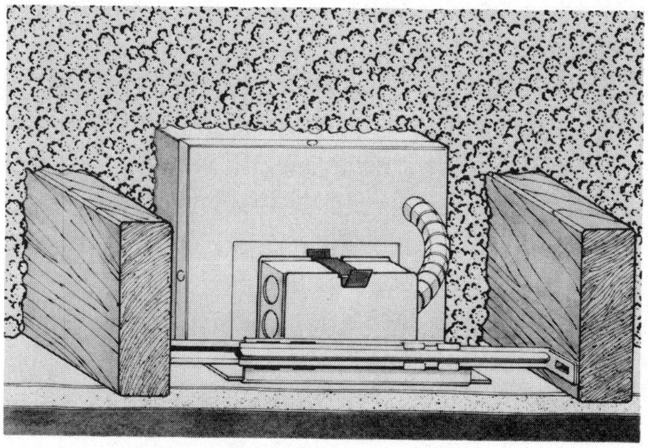	MARKED "TYPE IC." USUALLY LOW-WATTAGE FOR IN-STALLATION IN DIRECT CONTACT WITH INSULATION. MAY BE BLANKETED WITH INSULATION. RECOMMENDED FOR HOMES. THESE FIXTURES OPERATE UNDER 90°C WHEN COVERED WITH INSULATION. NO THREE-INCH SPAC-ING OF INSULATION REQUIRED. INTEGRAL THERMAL PROTECTION NOT REQUIRED. MAY ALSO BE USED IN NON-INSULATED AREAS. MANY OF THESE FIXTURES HAVE "DOUBLE-WALL" CONSTRUCTION.
NONINSULATED CEILINGS 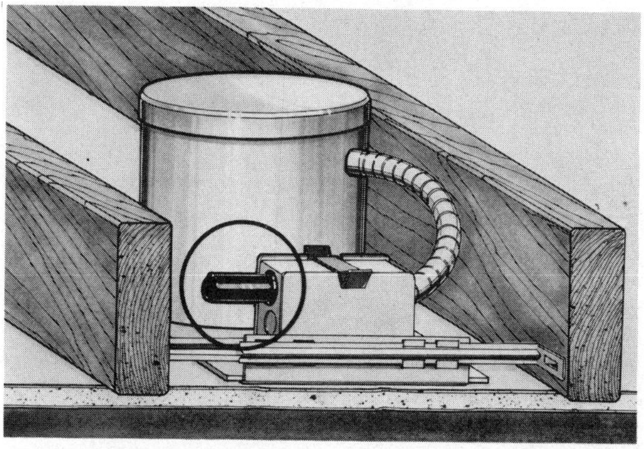	HAS BUILT-IN THERMAL PROTECTION. USE WHERE INSU-LATION IS NOT IN DIRECT CONTACT WITH FIXTURE. INSU-LATION MUST BE PULLED BACK AT LEAST THREE INCHES. OTHERWISE, THERMAL DEVICE WILL DEACTIVATE THE FIXTURE, CYCLING IT "ON—OFF—ON—OFF—ON—OFF" UN-TIL THE HEAT PROBLEM IS REMOVED. NOTE THERMAL PROTECTION IN CIRCLE.
SUSPENDED CEILINGS 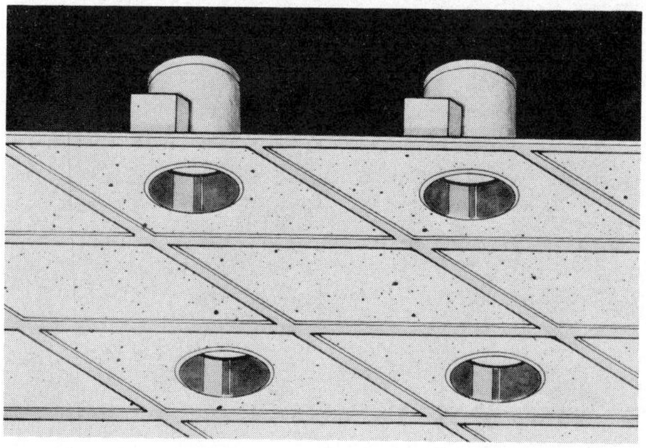	USE ONLY IN SUSPENDED, NONINSULATED CEILINGS. THERMAL PROTECTION NOT REQUIRED.

Fig. 15-8 Types of recessed fixtures.

box mounted on the side of the fixture so that the branch-circuit conductors can be run directly into the box and then connected to the conductors entering the fixture.

Additional wiring is unnecessary with these prewired fixtures, figure 15-11. Note that *Section 410-11* states that branch-circuit wiring shall not be passed through an outlet box that is an integral part of an incandescent fixture unless the fixture is identified for through wiring.

If a recessed fixture is not prewired, the electrician must check the fixture for a label indicating that it is necessary to use a conductor which has an insulation temperature rating of more than 60°C (140°F).

Recessed fixtures are inserted into the rough-in opening and fastened in place by various devices. One type of support and fastening method for recessed fixtures is shown in figure 15-12. The flag hanger remains against the fixture until the screw is turned. The flag then swings into position and hooks over the support rail as the screw is tightened.

Thermal Protection (*Section 410-65*). Recessed incandescent fixtures installed after April 1, 1982,

must have some type of integral thermal protection, *NEC Section 410-65*. Thermal protection will be provided in or on the fixture. The purpose of thermal protection is to open the circuit supplying the fixture should the temperature exceed the limit for the specific fixture.

Thermal protection is not required where the recessed fixture is identified for use and installation in poured concrete, or if the recessed fixture is identified for installations where the thermal insulation will come into direct contact with the fixture. Fixtures must be so marked by the manufacturer in order to obtain UL listing.

Wiring

It was stated previously that it is very important to provide an exact rough-in for surface-mounted fixtures. *Section 410-14(b)* of the *NEC*

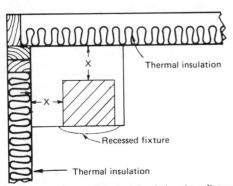

X distance at least 3 inches. Insulation above fixture must not trap heat; fixture and insulation must be installed to permit free air circulation unless the fixture is identified for installation within thermal insulation.

Fig. 15-9 Clearances for recessed lighting fixture installed near thermal insulation. *Section 410-66.*

Fig. 15-10 Boxlike device prevents insulation from coming into contact with fixture. *Section 410-66.*

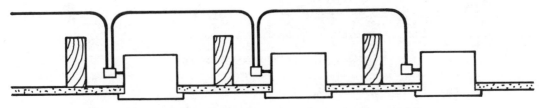

Fig. 15-11 Installation permissible only with prewired recessed fixtures.

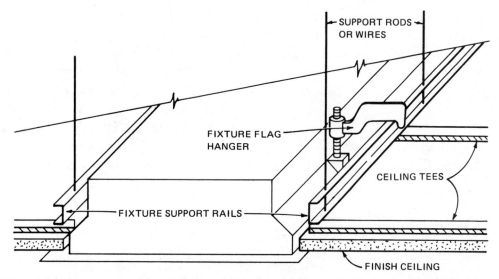

Fig. 15-12 Recessed fixture supported with flag hanger and support rails. *Section 410-16(c).*

emphasizes this statement by requiring that the lighting outlet be accessible for inspection (without removing the surface-mounted fixture). The installation of the outlet meets the requirements of *Section 410-14(b)* if the lighting outlet is located so that the large opening in the back of the fixture can be placed over it, figure 15-13.

To meet the requirements of *Section 410-31*, branch-circuit conductors with a rating of 90°C may be used to connect fixtures. However, these conductors must be of the single branch circuit supplying the fixtures. Therefore, all of the conductors of a multiwire branch circuit can be installed as long as these conductors are the grounded and ungrounded conductors of a single system. For example, since the beauty salon in the commercial building has a three-phase, four-wire supply, the neutral and three hot wires, one on each phase, can be installed in a fixture if approved as a raceway. Although this method of connecting fixtures is not practical in the beauty salon, it is suited to the installation of a long continuous row of fixtures, figure 15-14.

Loading Allowance

The branch circuits must be sized when the lighting layout is made. The actual fixture load may vary slightly for fixtures from different manufacturers. Therefore, it is customary to use a value that is larger than the typical load. At the same time, this selection will simplify the calculations.

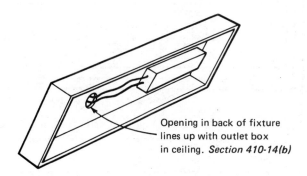

Opening in back of fixture lines up with outlet box in ceiling. *Section 410-14(b)*

Fig. 15-13

The typical loading values assumed for the fixtures used in the commercial building are shown in figure 15-15. The following sections describe each style of luminaire listed in figure 15-15.

SURFACE-MOUNTED LUMINAIRES

Style A

The Style A luminaire is a popular fluorescent type which features a diffuser extending up the sides of the luminaire, figure 15-16. This type of luminaire provides good ceiling lighting which is particularly important for low ceilings. The major disadvantage of the Style A luminaire is the difficulty of finding replacement diffusers several years after the luminaire is installed. The diffuser is usually available in either an acrylic or polystyrene material. Although a polystyrene diffuser is less expensive, it will yellow as it ages and thus will reduce the light output.　In addition, the general appearance of the luminaire is affected as the diffuser changes color.　Diffusers made from the acrylic material do not yellow with age.

Style B

The Style B luminaire, another popular style, has solid metal sides, figure 15-17.　Since this type of luminaire does not light the ceiling to any appreciable degree, the Style B luminaire is generally used where the ceilings are higher or in classrooms or offices where the bright sides of the Style A luminaire may be objectionable to people who must look at them for long periods of time.　The opal glass of the Style B luminaire provides a soft diffusion of the light, but any flat diffuser may be used.　Glass is easily cleaned and does not experience

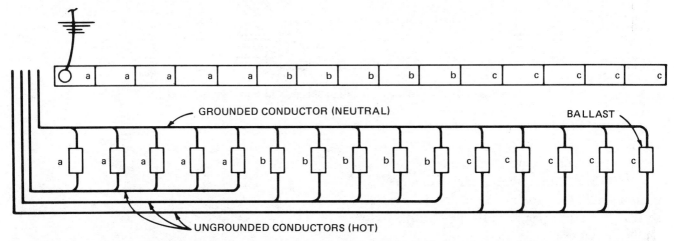

Fig. 15-14　A single branch circuit consisting of one grounded and three ungrounded conductors of the same system supplying a continuous row of approved fixtures.　See *NEC Section 410-31*.　Fixtures can be connected alternately:　a-b-c-a-b-c-a-b-c.

Fixture Mark*	Lamps	Typical Volt-amperes	Loading Allowance**
A	4-F40CW	192	200
B	4-F20T12CW	116	120
C	2-F40CW	96	100
D	2-F40CW	96	100
E1	1-F40WW	54	60
E2	8-R20	400	400
E3	1-R30	75	75
F	4-F40CW	192	200
G	2-F40CW/U	96	100
H	2-F40CW	96	100
I	2-F40CW	96	100
J	1-150A23/99	150	150
K	1-H37-5KC/DX	315	315
L	2-F40CW	96	100
M1	1-150R/SP	150	150
M2	2-150R/SP	300	300

*The various styles of luminaires, as indicated by these letter designations, are described in the following paragraphs.

**These are arbitrary values and are given only to simplify the calculations when exact values are not known.

Fig. 15-15　Loading values for luminaires.

the same aging problems encountered by the plastic materials.

Style C

The Style C luminaire is used where the possible contamination of the area is an important consideration, such as in areas where food is prepared. Style C luminaires are suitable for use in bakeries, restaurants, slaughterhouses, meat markets, and food-packaging plants. The clear acrylic diffuser of this luminaire protects the area in the event of a broken lamp. At the same time, the interior of the luminaire is kept dry and free from dirt or dust, figure 15-18

Style D

Luminous ceiling systems are used where a high level of diffuse light is required. The Style D system consists of fluorescent light strips (which may be high-output or very high-output rapid-start lamps), and a ceiling suspended 18 inches (457 mm) or more below the lamps, figure 15-19. The ceiling may be of any translucent material but usually consists of 2' x 2' (610 mm x 610 mm) or 2' x 4' (0.6 m x 1.22 m) panels which are easily removable for cleaning and lamp replacement.

Style E

Style E luminaires are used in the commercial building to light the show windows, figure 15-20.

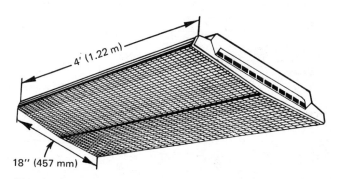

Fig. 15-16 Style A. A shallow, surface-mounted fluorescent luminaire, 4' (1.22 m) long and 18" (457 mm) wide with a wraparound acrylic diffuser and equipped for 4-F40CW rapid-start lamps.

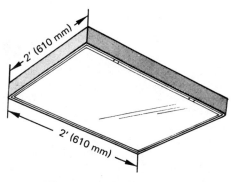

Fig. 15-17 Style B. A surface-mounted fluorescent luminaire, 2' (610 mm) square with solid sides, an opal glass diffuser, and equipped for 4-F20T12/CW lamps for trigger start.

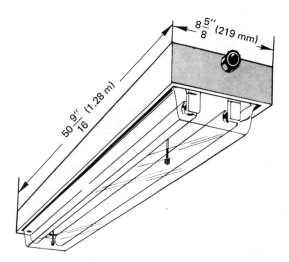

Fig. 15-18 Style C. A surface-mounted fluorescent luminaire 50 9/16" (1.28 m) long and 8 5/8" (219 mm) wide. Designed to prevent contamination of the area by having a totally enclosed and gasketed clear acrylic diffuser, equipped for 2-F40CW rapid-start lamps.

The Style E1 fixture uses a fluorescent lamp and is selected for the bakery because its lower heat output (compared to the incandescent lamps) will not damage the baked goods on display. The Style E2 luminaires uses incandescent lamps which are available in different colors and sizes. The Style E3 luminaire has a single incandescent lamp and is used to accent a display.

Style I

The Style I luminaire shown in figure 15-21 is designed to light corridors, the narrow areas between storage shelves, and other long, narrow spaces. Style I luminaires are available in slightly different styles from various manufacturers.

Style K

The Style K luminaire is used on vertical exterior walls where it provides a light pattern that covers a large area. The Style K luminaire uses a mercury lamp or other HID source, figure 15-22. This fixture provides reliable security lighting around a building. The Style K luminaire is completely weatherproof and is equipped with a photoelectric cell to turn on the light automatically at sunset.

Style L

The Style L luminaire is of open construction, as shown in figure 15-23. This type is generally used in storage areas and other locations where it is not necessary to shield the lamps. Style

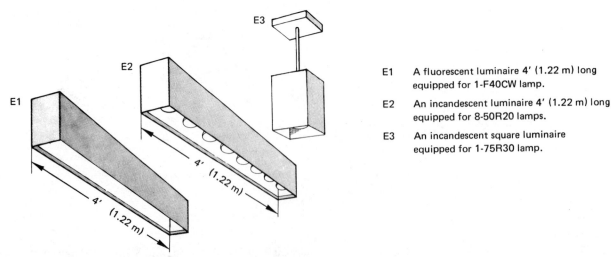

Fig. 15-19 Style D. A fluorescent light strip 8′ (2.44 m) long, equipped with 2-F40CW rapid-start lamps in tandem. Installed as a part of a luminous ceiling system with 2′ x 2′ (610 mm x 610 mm) panels with 1/2″ x 1/2″ (12.7 mm x 12.7 mm) cell size.

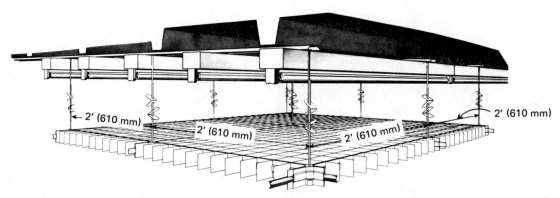

E1 A fluorescent luminaire 4′ (1.22 m) long equipped for 1-F40CW lamp.

E2 An incandescent luminaire 4′ (1.22 m) long equipped for 8-50R20 lamps.

E3 An incandescent square luminaire equipped for 1-75R30 lamp.

Fig. 15-20 Style E. Matching linear and square luminaires 6″ (152 mm) deep and 4 1/4″ (108 mm) wide, all with open bottoms.

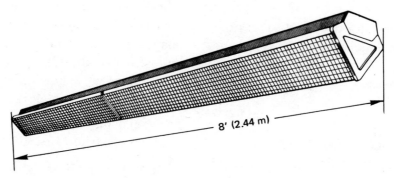

Fig. 15-21 Style I. A surface-mounted luminaire designed for corridor lighting with 3/8″ (9.5 mm) cube, V-shaped diffuser 8′ (2.44 m) long, equipped for 2-F40CW rapid-start lamps in tandem.

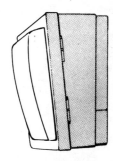

Fig. 15-22 Style K. A mercury luminaire for vertical surface mounting with an aluminum reflector, a prismatic refractor, and equipped for 1-H37-5KC/DX lamp.

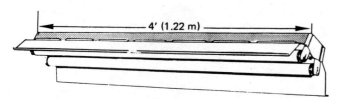

Fig. 15-23 Style L. An industrial-type fluorescent luminaire, 4′ (1.22 m) long with a steel white enameled reflector, equipped for 2-F40CW lamps.

MI   A ONE-LAMP UNIT EQUIPPED FOR 1-150 R/SP
M2   A TWO-LAMP UNIT EQUIPPED FOR 2-150 R/SP

Fig. 15-24 Style M. An adjustable shade incandescent canopy unit.

L luminaires often are suspended from the ceiling on chains. In this type of installation a rubber cord runs from the luminaire to a receptacle outlet on the ceiling.

Style M

The Style M luminaire is used to focus the light on a specific object, figure 15-24. This type of fixture can be swiveled through 360°. A wide selection in the sizes and designs of Style M lamps provides a variety of choices for accent lighting.

RECESSED-MOUNTED LUMINAIRES

Style F

The Style F recessed luminaire is an attractive lighting fixture for certain installations. This type of luminaire is available with many styles of diffusers, some of which extend below the ceiling surface to provide a small amount of light on the ceiling. The prismatic diffuser specified provides an even distribution of light with high efficiency, figure 15-25.

Style G

The Style G luminaire is basically the same as the Style F luminaire. However, this fixture measures 2 ft. by 2 ft. (610 mm x 610 mm) and uses a 40-watt, U-shaped lamp, figure 15-26.

Style H

The Style H luminaire, figure 15-27, is the same as the Style F fixture except that it is 1 ft. x

2 ft. (305 mm x 610 mm) and has only two lamps. The fixture is shown with a cube-type diffuser which permits air circulation around the lamps and does not trap dirt (a problem which commonly occurs with a solid diffuser).

Style J

The Style J luminaire, figure 15-28, is a recessed incandescent fixture which can be installed in the opening left by removing or omitting a single one-foot square ceiling tile. The convex diffuser extends below the ceiling and provides a wide distribution of light. When equipped with a lamp having a long life, this luminaire makes a good entry light.

LOCATION OF FIXTURES IN CLOTHES CLOSETS

Clothing, boxes, and other material normally stored in clothes closets are a potential fire hazard.

These items may ignite on contact with the hot surface of an exposed light bulb. The bulb, in turn, may shatter and spray hot sparks and hot glass onto other combustible materials. *NEC Section 410-8* gives the following rules for the installation of lighting fixtures in clothes closets.

- Pendants shall *not* be installed in clothes closets, figure 15-29.
- Surface-mounted clothes closet fixtures must be placed so that there is an 18-inch (145 mm) clearance between the fixtures and combustible material storage areas, figure 15-31.;

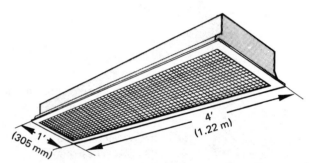

Fig. 15-27 Style H. An identical luminaire to F except 1′ (305 mm) wide and 4′ (1.22 m) long and equipped for 2-F40CW rapid-start lamps.

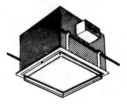

Fig. 15-28 Style J. A recessed incandescent luminaire 12″ (305 mm) square and 6″ (152 mm) deep with a convex glass diffuser for wide distribution; equipped for 1-150/99CL vertically mounted lamp.

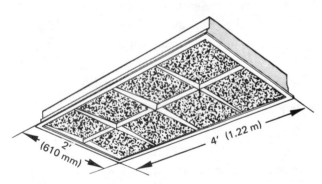

Fig. 15-25 Style F. A recessed troffer fluorescent luminaire 2′ (610 mm) wide and 4′ (1.22 m) long, designed for installation in an acoustical tile on plaster ceiling, with a prismatic diffuser and equipped for 4-F40CW rapid-start lamps.

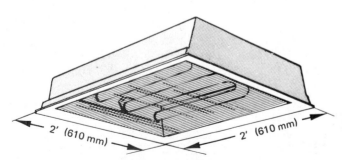

Fig. 15-26 Style G. An identical luminaire to F except 2′ (610 mm) square and equipped for 2-F40/CW/U lamps.

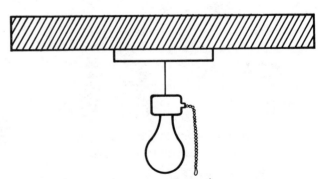

Fig. 15-29 Pendants not permitted in closets.

• Flush, recessed fixtures having a solid lens, or ceiling-mounted fluorescent fixtures, must be placed so that there is at least a 6-inch (152-mm) clearance between the fixtures and the combustible material storage area, figure 15-32.

Fig. 15-30 Recessed closet light with pull-chain switch for use where insulation will cover fixture and where separate wall switch is not provided.

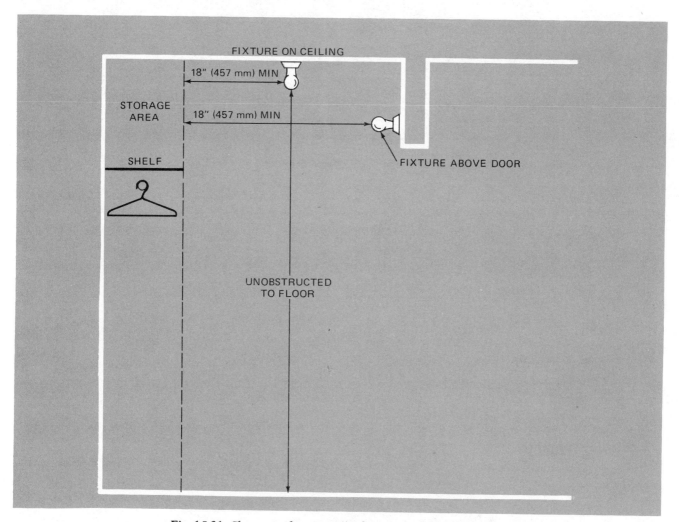

Fig. 15-31 Clearances for mounting fixtures in clothes closets.

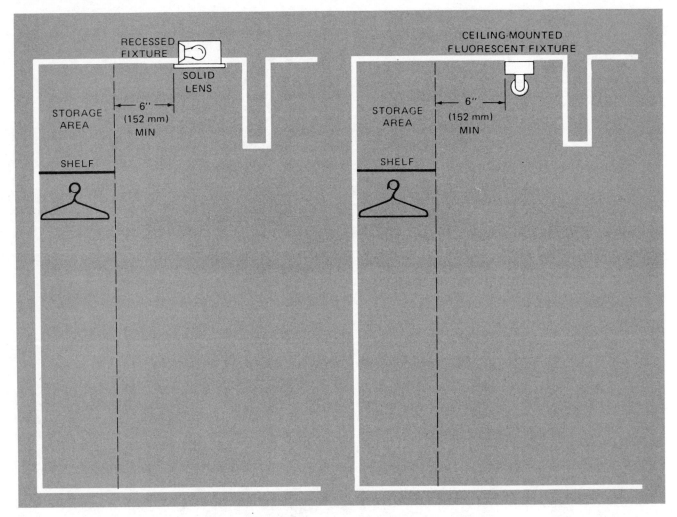

Fig. 15-32 Clearances for mounting fixtures in clothes closets.

REVIEW

Note: Refer to the *National Electrical Code* or the plans as necessary.

1. A lighting fixture must be supported from a _____ of the building.

2. Flexible metal raceway to a recessed fixture must be more than _____ feet (_____ m) and less than _____ feet (_____ m) long.

3. _____ neutral and _____ phase wires of a three-phase, four-wire system may be installed in a continuous row if the wires have a _____ °C rating.

4. Using a luminaire with mounting holes as shown in figure 15-6, draw a floor plan indicating the exact location of the center of the lighting outlets and support rods. Two 8-ft (2.44 m) and two 4-ft (1.22 m) luminaires are to be installed in two 12-ft rows.

5. The loading allowance for luminaires in the beauty salon is _____ volt-amperes.

6. In the beauty salon, the center of the first fixture is _____ feet _____ inches from the west wall.

7. Mercury luminaires are not used in corridor and office lighting because _____ _____ .

8. For an installation in a meat market, a luminaire similar to Style _____ should be used.

9. Bookshelving in a library is best lighted by using a luminaire similar to Style _____ .

10. In a clothes closet, a clearance of _____ inches must be maintained between a recessed lighting fixture and adjacent combustible material, such as wood joists.

11. What are the requirements for installing a recessed lighting fixture that does not have an approved junction box furnished with the fixture?

_____
_____
_____
_____
_____
_____
_____

UNIT 16

Special Circuits (Owner's Circuits)

OBJECTIVES

After completing the study of this unit, the student will be able to

- diagram the proper connections for photocells and timer-controlled lights.
- list and describe the main parts of an electric boiler control.
- connect the sump pump.

Each occupant of the commercial building is responsible for electric power used within that area. It is the owner's responsibility to provide the power to light the public areas and to provide heat for the entire building. The circuits supplying the power to those areas and devices for which the building is responsible will be called the owner's circuits in this unit.

LOADING SCHEDULE

The loading schedule for the owner's circuits indicates that these circuits use the greatest proportion of the electric power metered in the commercial building, figure 16-1. The loading schedule can be divided into three parts:

- the boiler feed which has a separate 800-ampere main switch in the service equipment.
- the emergency power system including the equipment and devices that must continue to operate if the utility power fails.
- the remaining circuits that do not require connection to the emergency system.

LIGHTING CIRCUITS

Those lighting circuits which are considered to be the owner's responsibility can be divided into

four groups according to the method used to control the circuit:

- continuous operation.
- manual control.
- timed control.
- photocell control.

Continuous Operation

The luminaires installed at the top of each of the stairways to the second floor of the commercial building are in continuous operation. Since the stairways have no windows, it is necessary to provide artificial light even in the daytime. The power to these luminaires is supplied directly from the panelboard.

Manual Control

A conventional lighting control system supplies the remaining second floor corridor lights as well as several other miscellaneous lights. All of the lights on manual control can be turned on as needed.

Timed Control

The lights at the entrance to the commercial building are controlled by a time clock located

OWNER'S LOADING SCHEDULE

Loading	NEC Article	Load in Watts		
		NEC	Actual	Design
Boiler Load				
Rated input		225,000	225,000	
25% of rated input	424-3(b)	56,250		
		281,250		281,250
General Lighting Load				
8 Style I luminaires			800	
7 Style J luminaires			1,050	
3 Style K luminaires @ 315 volt-amperes			945	
6 Style L luminaires			600	
			3,395	3,395
Motor Load				
Sump pump 208 volts × 5.4 amperes		1,123	1,123	
25% of sump pump		281		
5 circulating pumps 1/6 hp 120 volts × 4.4 amperes		2,640	2,640	
25% of circulating pumps	424-3(b)	660		
		4,704		4,704
Other Load				
Receptacle outlets				
2 special for telephone equipment		2,500	2,500	
4 outlets @ 180 VA/outlet		720		
		3,220		3,220
Total				292,569

Fig. 16-1

near the owner's panel in the boiler room. The circuit is connected as shown in figure 16-2. When the time clock is first installed, it must be adjusted to the correct time. Thereafter, it automatically controls the lights. The clock motor is connected to the emergency panel so that the correct time is maintained in the event that the utility power fails.

Photocell Control

The three lights located on the exterior of the building are controlled by individual photocells, figure 16-3. The photocell control consists of a light-sensitive photocell and an amplifier which increases the photocell signal until it is sufficient to operate a relay which controls the light. The neutral of the circuit must be connected to the photocell to provide power for the amplifier and relay.

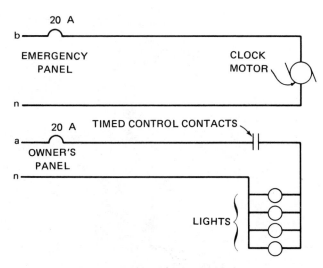

Fig. 16-2 Timed control of lights.

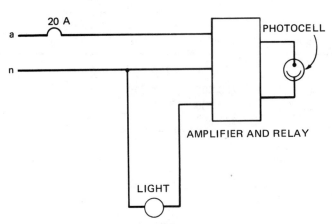

Fig. 16-3 Photocell control of light.

Fig. 16-4 Manual motor controller for sump pump.

Fig. 16-5 Float switch.

SUMP PUMP CONTROL

The sump pump is used to remove water entering the building because of sewage line backups, water main breakage, minor flooding due to natural causes, or plumbing system damage within the building. Since the sump pump is a critical item, it is connected to the emergency panel. The pump motor is protected by a manual motor starter, figure 16-4, and is controlled by a float switch, figure 16-5. When the water in the sump rises, a float is lifted which mechanically completes the circuit to the motor and starts the pump, figure 16-6. When the water level falls, the pump shuts off.

One type of sump pump that is available commercially has a start-stop operation due to water pressure against a neoprene gasket. This gasket, in turn, pushes against a small integral switch within the submersible pump. This type of sump pump does not use a float.

BOILER CONTROL

An electrically heated boiler supplies heat to all areas of the commercial building. The boiler has a full-load rating of 225 kW. As purchased, the boiler is completely wired except for the external control wiring, figure 16-7. A heat sensor plus a remote bulb, figure 16-8, is mounted so that the bulb is on the outside of the building. The bulb must be mounted where it will not receive direct sunlight and must be spaced at least 1/2 inch from the brick wall. If these mounting instructions are not followed, the bulb will give inaccurate readings. The heat sensor is adjusted so that it closes when the outside temperature falls below a set point, usually 18°C (65°F). The sensor remains closed until the temperature increases to a value above the differential setting, approximately 21°C (70°F). Once the outside temperature causes the heat sensor to close, the boiler automatically maintains a constant water temperature and is always ready to deliver heat to any zone in the building that requires heat.

A typical heating control circuit is shown in figure 16-9.

Figure 16-10 is a schematic drawing of a typical connection scheme for the boiler heating elements.

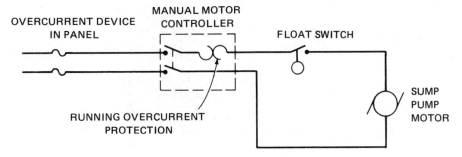

OVERCURRENT DEVICE
IN PANEL

MANUAL MOTOR
CONTROLLER

FLOAT SWITCH

RUNNING OVERCURRENT
PROTECTION

SUMP
PUMP
MOTOR

Fig. 16-6 Sump pump control.

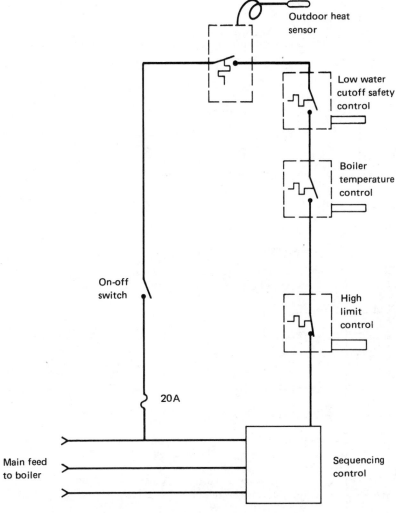

Outdoor heat
sensor

Low water
cutoff safety
control

Boiler
temperature
control

On-off
switch

High
limit
control

20A

Main feed
to boiler

Sequencing
control

Fig. 16-7 Boiler control diagram.

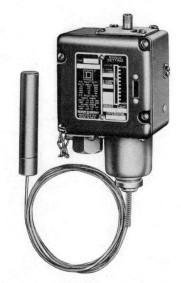

Fig. 16-8 Heat sensor with remote mounting
bulb.

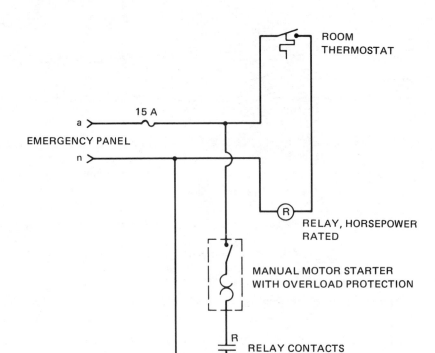

Fig. 16-9 Heating control circuit.

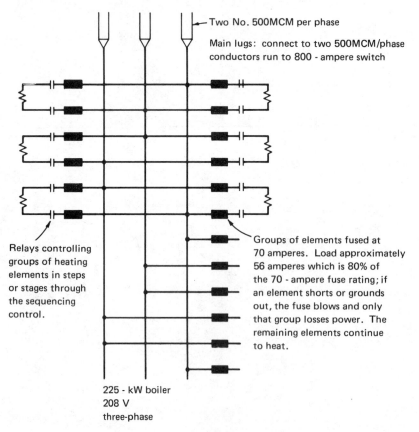

Fig. 16-10 Boiler heating elements.

The following points summarize the wiring requirements for a boiler.

- A disconnect means must be installed to disconnect the ungrounded supply conductors.

- The disconnect must be in sight or capable of being locked in the open position.

- The circuit conductor sizes are to be based on 125% of the rated load.

- The overcurrent protective devices are to be sized at 125% of the rated load.

Article 424 of the *NEC* covers the Code rules for fixed electric space heating equipment.

It should be noted that the boiler for the commercial building is not sized for a particular heat loss; its selection is based on certain electrical requirements. Each of the building occupants has a heating thermostat located within the particular area. The thermostat operates a relay, figure 16-11, which controls a circulating pump, figure 16-12, in the hot water piping system serving the area. The circuit to the circulating pump is supplied from the emergency panel so that the water will continue to circulate if the utility power fails. In subfreezing weather, the continuous circulation prevents the freezing of the boiler water; such freezing can damage the boiler and the piping system.

Fig. 16-11 Relay in NEMA 1 enclosure.

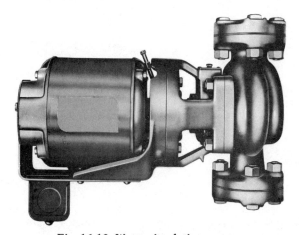

Fig. 16-12 Water circulating pump.

REVIEW

Note: Refer to the *National Electrical Code* or the plans as necessary.

1. What sizes of conduit(s) and wire are required to supply electric power to the boiler?
   _____

2. A _____-ampere overload protective device should be installed in the manual motor controller for the sump pump. (Show calculations.)
   _____

3. There are _____ circulating pumps.

4. There is a (an) _____ volt-ampere load connected to the emergency panel.

5. The outside heat sensor should not be installed where the _____ or the _____ can affect its reading.

6. Find the maximum trip rating in amperes for an inverse time circuit breaker used to provide short-circuit protection to a circulating pump motor branch circuit. (Show calculations.) _____
   _____
   _____

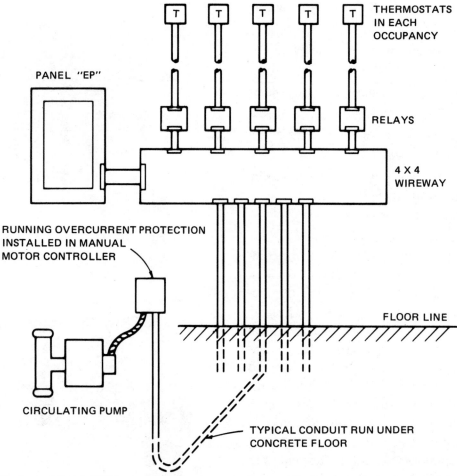

Fig. 16-13   Installation of Panel "EP" relays, thermostats, conduits, wireway, and motor controller to connect circulating pumps for each occupant.

7. List four methods of controlling lighting circuits.

a. _____

b. _____

c. _____

d. _____

8. The disconnect means for the boiler must be located in sight of the boiler or _____

_____ .

9. Conductors supplying an electrically heated boiler must have an ampacity of at least _____ percent of the rated load as required by *NEC Section* _____ .

10. The sump pump is controlled by a(n) _____.

11. What is the actual current draw of the boiler when all elements are on? (Show calculations.)   _____

_____

_____

═══UNIT 17═══
Emergency Power Systems

─── OBJECTIVES ───

After completing the study of this unit, the student will be able to

• select and size an emergency power system.

• install an emergency power system.

Many state and local codes require that equipment be installed in public buildings to insure that electric power is provided automatically if the normal power source fails. The electrician should be aware of the special installation requirements of these emergency systems. *Article 700* of the *NEC* covers emergency system Code requirements.

SOURCES OF POWER (*NEC Article 700, Part B*)

Battery Powered Lighting [*Section 700-12(a)*]

When the need for emergency power is confined to a definite area, such as a stairway, and the power demand in this area is low, then self-contained battery powered units are a convenient and efficient means of providing power. In general, these units are wall-mounted and are connected to the normal source by permanent wiring methods. Under normal conditions, this regular source powers a regulated battery charger to keep the battery at full power. When the normal power fails, the circuit is completed automatically to one or more lamps which provide enough light to the area to permit its use, such as lighting a stairway sufficiently to allow people to leave the building. Battery powered units are commonly used in stairwells, hallways, shopping centers, supermarkets, and other public structures.

Central Battery Power

If the power demand is high (excluding the operation of large motors), central battery power systems are available. These systems usually operate at 32 volts and can service a large number of lamps. For example, central systems are available that can provide over 2,500 watts per hour.

SPECIAL SERVICE ARRANGEMENTS

If separate utility power sources are available *Section 700-12(d)*, figure 17-1, emergency power can be provided at a minimum cost. If protection is required to cover power failures within the building only, then a connection is made ahead of the main switch, figure 17-2. When an emergency circuit connection is made ahead of the main switch *Section 700-12(e)*, the requirements of *NEC Section 230-98* must be followed. In other words, all of the equipment must have an adequate interrupting rating for the available short-circuit current (Unit 18).

EMERGENCY GENERATOR SOURCE [*Section 700-12(b)*]

Generator sources may be used to supply emergency power. The selection of such a source for a specific installation involves a consideration of the following factors:

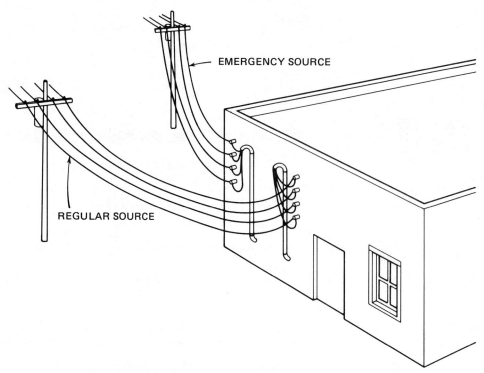

Fig. 17-1 Separate services. *Section 700-12(d).*

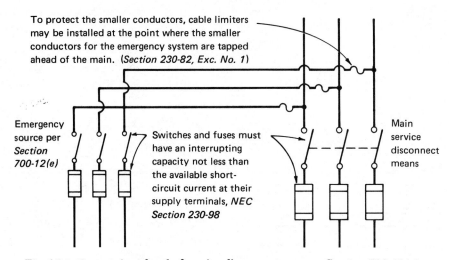

To protect the smaller conductors, cable limiters may be installed at the point where the smaller conductors for the emergency system are tapped ahead of the main. *(Section 230-82, Exc. No. 1)*

Emergency source per *Section 700-12(e)*

Switches and fuses must have an interrupting capacity not less than the available short-circuit current at their supply terminals, *NEC Section 230-98*

Main service disconnect means

Fig. 17-2 Connection ahead of service disconnect means. *Section 700-12(e).*

- the engine type.
- the generator capacity.
- the load transfer controls.

A typical generator for emergency use is shown in figure 17-3.

Engine Types and Fuels

The type of fuel to be used in the driving engine of a generator is an important consideration in the installation of the system. Fuels which may be used are LP gas, natural gas, gasoline, or diesel fuel. Factors affecting the selection of the fuel to be used include the availability of the fuel and local regulations governing the storage of the fuel. Natural gas and gasoline engines differ only in the method of supplying the fuel; therefore, in some installations, one of these fuels may be used as a standby alternative for the other fuel, figure 17-4.

An emergency power source which uses gasoline and/or natural gas usually has lower installation and operating costs than a diesel-powered source.

Fig. 17-3 Typical generator for emergency use.

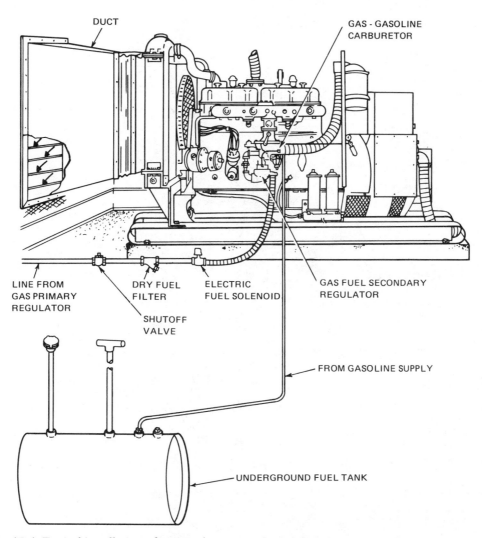

Fig. 17-4 Typical installation of a natural gas — gasoline engine-driven generator. *Section 700-12(b)*.

However, the problems of fuel storage can be a deciding factor in the selection of the type of emergency generator. Gasoline is not only dangerous, but becomes stale after a relatively short period of time and thus cannot be stored for long periods. If natural gas is used, the Btu content must be greater than 1,100 Btu per cubic foot. Diesel-powered generators require less maintenance and have a longer life. This type of diesel system is usually selected for installations having large power requirements since the initial costs closely approach the costs of systems using other fuel types.

Cooling

Smaller generator units are available with either air or liquid cooling systems. Units having a capacity greater than 15 kW generally use liquid cooling.

For air-cooled units, it is recommended that the heated air be exhausted to the outside. In addition, a provision should be made to bring in fresh air so that the room where the generator is installed can be kept from becoming excessively hot. Typical installations of air-cooled emergency generator systems are shown in figures 17-5 and 17-6.

Generator Voltage Characteristics

Generators having any required voltage characteristic are available. A critical factor in the selection of a generator for a particular application is that it must have the same voltage output as the normal building supply system. For the commercial building covered in this text, the generator selected provides 120/208-volt, three-phase, 60-Hz power.

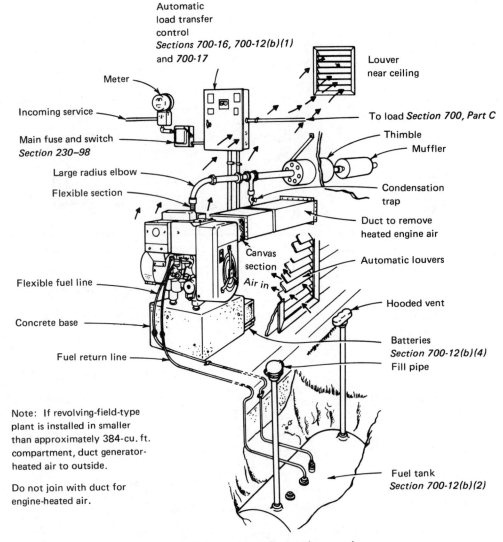

Fig. 17-5 Typical installation (pressure).

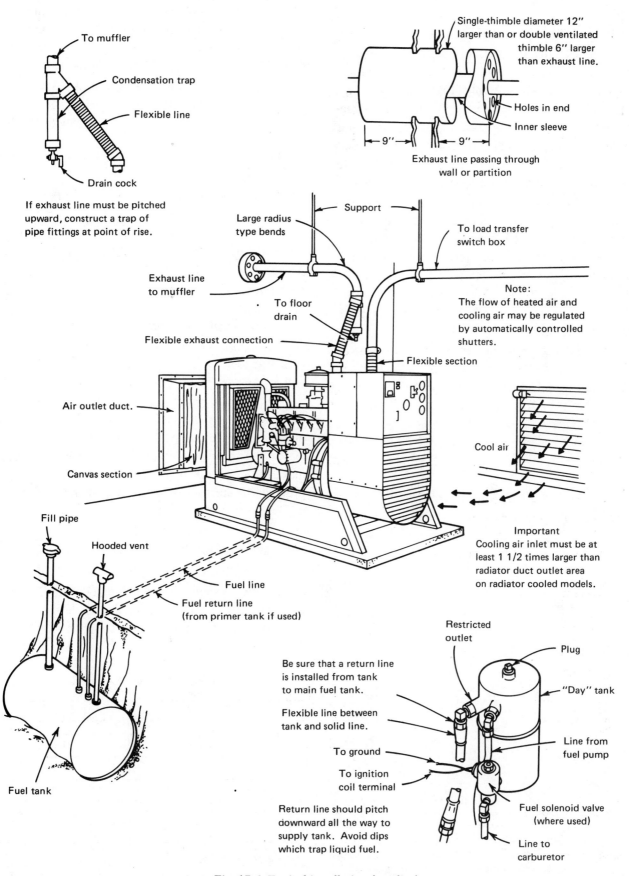

To muffler

Condensation trap

Flexible line

Drain cock

If exhaust line must be pitched upward, construct a trap of pipe fittings at point of rise.

Single-thimble diameter 12" larger than or double ventilated thimble 6" larger than exhaust line.

Holes in end

Inner sleeve

9"   9"

Exhaust line passing through wall or partition

Support

Large radius type bends

To load transfer switch box

Exhaust line to muffler

To floor drain

Note:
The flow of heated air and cooling air may be regulated by automatically controlled shutters.

Flexible exhaust connection

Flexible section

Air outlet duct.

Cool air

Canvas section

Important
Cooling air inlet must be at least 1 1/2 times larger than radiator duct outlet area on radiator cooled models.

Fill pipe

Hooded vent

Fuel line

Fuel return line (from primer tank if used)

Fuel tank

Restricted outlet

Be sure that a return line is installed from tank to main fuel tank.

Plug

"Day" tank

Flexible line between tank and solid line.

To ground

Line from fuel pump

To ignition coil terminal

Fuel solenoid valve (where used)

Return line should pitch downward all the way to supply tank. Avoid dips which trap liquid fuel.

Line to carburetor

Fig. 17-6 Typical installation (gasoline).

Capacity

It is an involved, but extremely important, task to determine the correct size of the engine-driven power system so that it has the minimum capacity necessary to supply the selected equipment. If the system is oversized, additional costs are involved in the installation, operation, and maintenance of the system. However, an undersized system may fail at the critical period when it is being relied upon to provide emergency power. To size the emergency system, it is necessary to determine initially all of the equipment to be supplied with emergency power. If motors are to be included in the emergency system, then it must be determined if all of the motors can be started at the same time. This information is essential to insure that the system has the capacity to provide the total starting inrush kVA required.

The starting kVA for a motor is equivalent to the locked rotor kVA of the motor. This value is determined by selecting the appropriate value from *NEC Table 430-7(b)* and then multiplying this value by the horsepower. The locked rotor kVA value is independent of the voltage characteristics of the motor. Thus, a 5-hp, code E motor requires a generator capacity of:

$$5 \text{ hp} \times 4.99 \text{ kVA/hp} = 24.95 \text{ kVA}$$

If two motors are to be started at the same time, the emergency power system must have the capacity to provide the sum of the starting kVA values for the two motors.

Table 430-7(b). Locked-Rotor Indicating Code Letters

Code Letter		Kilovolt-Amperes per Horsepower with Locked Rotor
A		0 — 3.14
B		3.15 — 3.54
C		3.55 — 3.99
D		4.0 — 4.49
E		4.5 — 4.99
F		5.0 — 5.59
G		5.6 — 6.29
H		6.3 — 7.09
J		7.1 — 7.99
K		8.0 — 8.99
L		9.0 — 9.99
M		10.0 — 11.19
N		11.2 — 12.49
P		12.5 — 13.99
R		14.0 — 15.99
S		16.0 — 17.99
T		18.0 — 19.99
U		20.0 — 22.39
V		22.4 — and up

For a single-phase motor rated at less than 1/2 hp, Table 17-1 lists the approximate kVA values that may be used if exact information is not available. The power system for the commercial building must supply the following maximum kVA.

Five 1/6 hp C-type motors
5 × 1.85 kVA = 9.25 kVA
One 1/2 hp Code L motor
1/2 × 9.99 kVA = 4.99 kVA
Lighting and receptacle
load from emergency
panel schedule = 4.42 kVA
Total   18.66 kVA

Thus, the generator unit selected must be able to supply this maximum kVA load as well as the continuous kVA requirement for the commercial building.

Continuous kVA requirement:
Lighting and receptacle load   4.420 kVA
Motor load   3.763 kVA
Total   8.183 kVA

A check of manufacturers' data shows that a 12-kVA unit is available having a 20-kVA maximum rating for motor starting purposes and a 12-kVA continuous rating. A smaller generator may be installed if provisions can be made to prevent the motors from starting at the same time.

Derangement Signals

According to *NEC Section 700-12*, a *derangement signal*, a signal device having both audible and visual alarms, must be installed outside the generator room in a location where it can be readily and

Table 17-1 Approximate kVA values

HP	Type	Locked Rotor kVA
1/6	C	1.85
1/6	S	2.15
1/4	C	2.50
1/4	S	2.55
1/3	C	3.0
1/3	S	3.25
C = Capacitor-start motor S = Split-phase motor		

regularly observed. The purposes of a device such as the one shown in figure 17-7 are to indicate: any malfunction in the generator unit, any load on the system, or the correct operation of a battery charger.

Automatic Transfer Equipment
(NEC Section 700-6)

If the main power source fails, equipment must be provided to start the engine of the emergency generator and transfer the supply connection from the regular source to the emergency source. These operations can be accomplished by a control panel such as the one shown in figure 17-8. A voltage-sensitive relay is connected to the main power source. This relay (transfer switch) activates the control cycle when the main source voltage fails, figure 17-9. The generator motor is started when the control cycle is activated. As soon as the motor reaches the correct speed, a set of contactors is energized to disconnect the load from its normal source and connect it to the generator output.

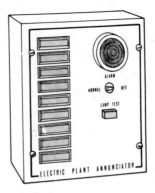
Fig. 17-7 Audible and visual signal alarm annunciator.

Fig. 17-8 Automatic transfer control panel.

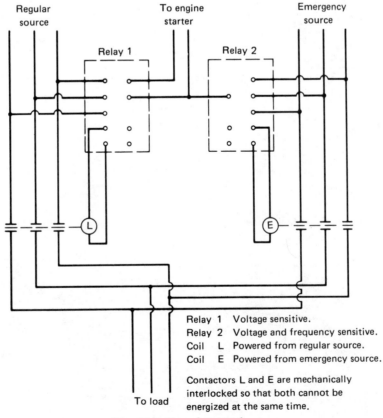

Relay 1 Voltage sensitive.
Relay 2 Voltage and frequency sensitive.
Coil L Powered from regular source.
Coil E Powered from emergency source.

Contactors L and E are mechanically interlocked so that both cannot be energized at the same time.

Fig. 17-9 Transfer switch.

Wiring

The branch-circuit wiring of emergency systems must be separated from the standard system except for the special conditions noted. Figure 17-10 shows a typical branch-circuit installation for an emergency system. Key-operated switches, figure 17-11, are installed to prevent unauthorized personnel from operating the lights.

Under certain conditions, emergency circuits are permitted to enter the same junction box as normal circuits, *Section 700-9, Exception.* Figure 17-12 shows an exit light which contains two lamps; one lamp connected to the normal circuit, and the other connected to the emergency circuit.

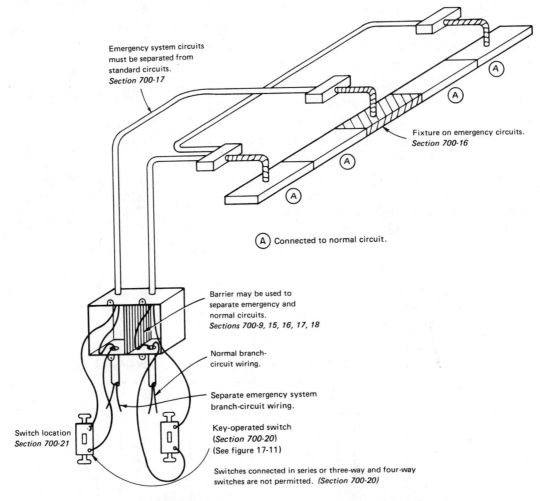

Emergency system circuits must be separated from standard circuits. *Section 700-17*

Fixture on emergency circuits. *Section 700-16*

Ⓐ Connected to normal circuit.

Barrier may be used to separate emergency and normal circuits. *Sections 700-9, 15, 16, 17, 18*

Normal branch-circuit wiring.

Separate emergency system branch-circuit wiring.

Switch location *Section 700-21*

Key-operated switch (*Section 700-20*) (See figure 17-11)

Switches connected in series or three-way and four-way switches are not permitted. *(Section 700-20)*

Fig. 17-10 Branch-circuit wiring for emergency systems and normal systems.

Fig. 17-11 Key-operated switch for emergency lighting control.

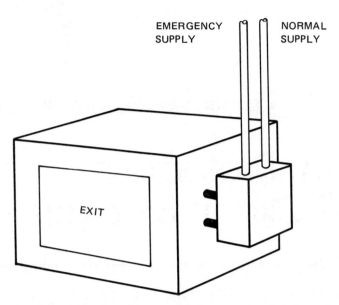

EMERGENCY SUPPLY

NORMAL SUPPLY

EXIT

Fig. 17-12 Both normal and emergency supply may enter the same junction box. *Section 700-9, Exception.*

REVIEW

Note: Refer to the *National Electrical Code* or the plans as necessary.

1. List three different methods of providing an emergency power source.
   1) _____ ,   2) _____ ,   3) _____

2. A 7 1/2-hp, three-phase, 230-volt, code F motor has a starting rating ranging between _____ kVA and _____ kVA. (Show calculations.)

   _____

   _____

3. If two 7 1/2-hp, three-phase, 230-volt, code F motors are to be started at the same time, the kVA capacity of the generator required is _____ kVA. (Show calculations.)

   _____

4. If two 7 1/2-hp, three-phase, 230-volt, code F motors are to be operated by a generator, the minimum size generator, in kVA capacity, that can be used to provide selective control for the motor is _____ kVA. (Show calculations.)

   _____

   _____

   _____

5. Name three conditions that can be indicated by a derangement signal device.
   1) _____ ,   2) _____ ,   3) _____

6. Emergency circuit wiring must be kept (together with) (separate from) normal circuit wiring. (Underline the correct answer.)

7. Three-way, four-way, and series-connected switches (are) (are not) permitted to control emergency lighting. (Underline the correct answer.)

UNIT 18

Overcurrent Protection: Fuses and Circuit Breakers

OBJECTIVES

After completing the study of this unit, the student will be able to

- list and identify the types, classes, and ratings of fuses and circuit breakers.
- describe the operation of fuses and circuit breakers.
- define interrupting capacity, short-circuit currents, $I^2 t$, I_p, rms, and current limitation.
- apply the *National Electrical Code* to the selection and installation of overcurrent protective devices.
- use the time-current characteristics curves and peak let-through charts.

Overcurrent protection is one of the most important components of an electrical system. The overcurrent device opens an electrical circuit whenever an overload or short circuit occurs. Overcurrent devices in an electrical circuit may be compared to pressure relief valves on a boiler. If dangerously high pressures develop within a boiler, the pressure relief valve opens to relieve the high pressure. In a similar manner, the overcurrent device in an electrical system also acts as a "safety valve."

The *National Electrical Code* requirements for overcurrent protection are contained in *Article 240* of the Code. *Section 240-1* states, in part, that overcurrent protection for conductors and equipment is provided to open the circuit if the current reaches a value that will cause an excessive or dangerous temperature in conductors or conductor insulation. See also *Sections 110-9* and *110-10* for requirements for interrupting capacity and protection against fault current.

Two types of overcurrent protective devices are commonly used: fuses and circuit breakers. The Underwriters Laboratories, Inc. (UL) and the National Electrical Manufacturers Association

(NEMA) establish standards for the ratings, types, classifications, and testing procedures for fuses and circuit breakers.

As indicated in *Section 240-6* of the *NEC*, the standard ampere ratings for fuses and nonadjustable circuit breakers are 15, 20, 25, 30, 35, 40, 45, 50, 60, 70, 80, 90, 100, 110, 125, 150, 175, 200, 225, 250, 300, 350, 400, 450, 500, 600, 700, 800, 1,000, 1,200, 1,600, 2,000, 2,500, 3,000, 4,000, 5,000 and 6,000. The *Exception* to *Section 240-6* lists additional standard ratings for fuses as 1, 3, 6, 10, and 601 amperes.

FUSES AND CIRCUIT BREAKERS

For general applications, the voltage rating, the continuous current rating, the interrupting rating, and the speed of response are factors that must be considered when selecting the proper fuses and circuit breakers.

Voltage Rating

According to *NEC Section 110-9*, the voltage rating of a fuse or circuit breaker should be equal to or greater than the voltage of the circuit in which the fuse or circuit breaker is to be used.

Continuous Current Rating

The continuous current rating of a fuse or circuit breaker should be equal to or slightly greater than the rating of the circuit or equipment it is to protect. For example, for a No. 8 TW wire with an ampacity of 40 amperes, a 40-ampere fuse or breaker is selected as the protective device. The rating of the fuse or circuit breaker may be greater than the current rating of the circuit only in special applications in which the connected equipment has unusual characteristics, such as for motors and transformers having high-inrush loads.

When the ampacity of a conductor does not match the amperage rating of a standard fuse or circuit breaker, the *NEC* permits the use of the next standard size fuse or circuit breaker. However, when the rating of the fuse or circuit breaker exceeds 800 amperes, then the conductor ampacity must be equal to or less than the rating of the fuse or circuit breaker, *Section 240-3, Exception No. 1.*

Interrupting Rating

The interrupting rating is a measure of the ability of a fuse or circuit breaker to safely open an electrical circuit under fault conditions, such as overload currents, short-circuit currents and ground-fault currents, without destroying itself. In other words, the interrupting rating of an overcurrent device is the maximum short-circuit current that it can interrupt safely at its rated voltage (see *NEC Section 110-9*).

Overload currents have the following characteristics:

- they are greater than the normal current flow.

- they are contained within the normal conducting current path.

- if allowed to continue, they will cause overheating of the equipment, conductors, and the insulation of the conductors.

Short-circuit and ground-fault currents, which flow outside of the normal current paths, can cause conductors to overheat. In addition, mechanical damage to equipment can occur as a result of the

magnetic forces of the large current flow and arcing. Some short-circuit and ground-fault currents may be no larger than the normal load current, or they may be thousands of times larger than the normal load current.

The terms *interrupting capacity* and *interrupting rating* are used interchangeably in the electrical industry.

Speed of Response

The time required for the fusible element of a fuse to open varies inversely with the magnitude of the current that flows through the fuse. In other words, as the current increases, the time required for the fuse to open decreases. The time-current characteristic of a fuse depends upon its rating and type. A circuit breaker also has a time-current characteristic. For the circuit breaker, however, there is a point at which the opening time cannot be reduced further due to the inertia of the moving parts within the breaker. The time-current characteristic of a fuse or circuit breaker should be selected to match the connected load of the circuit to be protected.

TYPES OF FUSES

Dual-element, Time-delay Fuse

The dual-element, time-delay fuse, figure 18-1, provides a time delay in the low-overload range to eliminate unnecessary opening of the circuit because of harmless overloads. However, this type of fuse is extremely responsive in opening on short circuits. This fuse type has two fusible elements connected

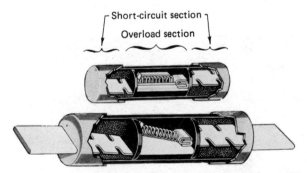

Fig. 18-1 Cutaway view of a Fusetron dual-element fuse. On overloads, the spring-loaded trigger assembly in the center chamber opens the fuse. The copper elements in both end chambers open when a short circuit occurs.

in series. Depending upon the magnitude of the current flow, one element or the other will open. The thermal cutout element is designed to open when the current reaches a value of approximately 500% of the fuse rating. The short-circuit element opens when a short circuit or heavy overloads occur. That is, the element opens at current values of approximately 500% or more of the fuse rating.

The thermal element is also designed to open at approximately 140°C (284°F), as well as on damaging overloads. In addition, the thermal element will open whenever a loose connection or a poor contact in the fuseholder causes heat to develop. As a result, the fuse also offers thermal protection to the equipment in which it is installed.

Dual-element fuses are suitable for use on motor circuits and other circuits having high-inrush characteristics. This type of fuse can be used as well for mains, feeders, subfeeders, and branch circuits. Dual-element fuses may be used to provide back-up protection for circuit breakers, bus duct, and other circuit components that lack an adequate interrupting rating, bracing, or withstand rating (covered later in this unit).

Dual-element fuses used on single-motor branch circuits generally are sized not to exceed 125% of the full-load running current of the motor. *NEC Table 430-152* shows that the maximum size permitted for a dual-element, time-delay fuse is 175% of the full-load current (FLA) of the motor. An exception to the 175% rating is given in *Section 430-52(b)*: a maximum rating of 225% of the motor FLA is allowed when the 175% value is not sufficient to provide protection for the starting current of the motor. It is possible to size a dual-element fuse so that it more nearly approximates the nameplate rating of the motor. Some engineers specify that dual-element fuses are to be sized at not over 110% of the full-load rating of the motor. Thus, the requirements of *NEC Article 430, Part C (Motor and Branch-circuit Running Overcurrent Protection)* and *Part D (Motor Branch-circuit Short-circuit Protection)*, can be met by properly sizing these fuses.

Dual-element, Time-delay, Current-limiting Fuses

The dual-element, time-delay, current-limiting fuse, figure 18-2, operates in the same manner as the dual-element, time-delay fuse. The only difference between the fuses is that this fuse has a faster response in the short-circuit range and thus is more current limiting. The short-circuit element in the current-limiting fuse usually is silver with a quartz sand arc-quenching filler.

Current-limiting Fuses

The standard current-limiting fuse, figure 18-3, has an extremely fast response in both the low-overload and short-circuit ranges. When compared to other types of fuses, this type of fuse has the lowest energy let-through values. Current-limiting fuses are used to protect mains, feeders and subfeeders, circuit breakers, bus duct, switchboards, and other circuit components that lack an adequate interrupting rating, bracing, or withstand rating.

In general, current-limiting fuse elements consist of silver links surrounded by a quartz sand arc-quenching filler.

Current-limiting fuses may be selected and loaded at 100% of their rating for lighting circuit loads. However, it is a good design practice not to exceed 80% of the rating for feeder and circuit loading.

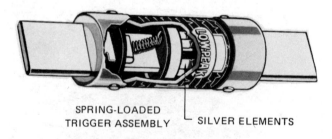

SPRING-LOADED
TRIGGER ASSEMBLY — SILVER ELEMENTS

Fig. 18-2 Cutaway view of low-peak, dual-element, current-limiting fuse. The spring-loaded trigger assembly in the center of the chamber opens on overloads, and the silver elements in the right-hand chamber open with a short circuit.

Fig. 18-3 Cutaway view of a Limitron current-limiting, single-element fuse.

When used on motor circuits or other circuits having high current-inrush characteristics, the current-limiting fuses must be sized at a much larger rating than the actual load. That is, for a motor with a full-load current rating of 10 amperes, a 30- or 40-ampere current-limiting fuse may be required to start the motor. In this case, the fuse is considered to be short-circuit protection only as required in *NEC Article 430*.

The definition of a current-limiting overcurrent protective device is given in *NEC Section 240-11*.

Current Limitation

Every piece of electrical equipment, including switches, motor controllers, conductors, bus duct, panels, and load centers has an ability to withstand a certain amount of electrical energy for a given amount of time before damage to the equipment occurs. Hence the term *withstand rating.*

Electrical equipment manufacturers conduct exhaustive tests to determine the withstand rating for their products. Equipment tested in this manner is marked with the class of fuse required (recommended). In addition, a maximum available fault-current rating is indicated for each piece of equipment.

The ability of a fuse to limit the let-through energy to a value less than the amount of energy that the electrical system is capable of delivering means that the equipment can be protected against fault-current values of high magnitudes. Equipment manufacturers specify fuses to minimize the damage that can occur in the event of a current fault. The value of the energy (in the form of heat) generated during a fault varies as the square of the rms- (root mean square) current multiplied by the time in seconds, I^2t. (I^2t is called "ampere squared seconds.") The magnetic forces acting on equipment vary as the square of the peak current, I_p^2.

Visual signs indicating that too much current was permitted to flow for too long a time include conductor insulation burning, melting and bending of bus bars, arcing damage, exploded overload elements in motor controllers, and welded contacts in controllers.

A current-limiting fuse (either a dual-element fuse or a straight current-limiting fuse) must be selected carefully. The fuse must have not only an adequate interrupting rating to clear a fault safely without damage to itself, but it also must be capable of limiting the let-through current (I_p) and the value of I^2t to the withstand rating of the equipment it is to protect.

The graph shown in figure 18-4 illustrates the current-limiting effect of fuses. The square of the area under the dashed line is energy (I^2t). I_p is the available peak short-circuit current that flows if the overcurrent device is not a current-limiting type. For a current-limiting fuse, the square of the shaded area of the graph represents the energy (I^2t), and the peak let-through current I_p. The melting time of the fuse element is t_m and the square of the shaded area corresponding to this time is the melting energy. The arcing time is shown as t_a; similarly, the square of the shaded area corresponding to this time is the arcing energy. The total clearing time, t_c, is the sum of the melting time and the arcing time. The square of the shaded area for time t_c is the total energy to which the circuit is subjected when the fuse has cleared. For the graph in figure 18-4, the area under the dashed line is six times greater than the shaded area. Since energy is equal to the area squared, then $6 \times 6 = 36$; that is, the circuit is subjected to 36 times as much energy when it is protected by noncurrent-limiting overcurrent devices.

Cartridge Fuses (*NEC Article 240, Part F*)

The *NEC* requirements governing cartridge fuses are contained in *Part F* of *Article 240*. According to the Code, all cartridge fuses must be marked to show:

- ampere rating.
- voltage rating.
- interrupting rating when over 10,000 amperes.
- current-limiting type, if applicable.
- trade name or name of manufacturer.

The Underwriters Laboratories requires that the fuse class be indicated as Class G, H, J, K, L, RK1, RK5, or T.

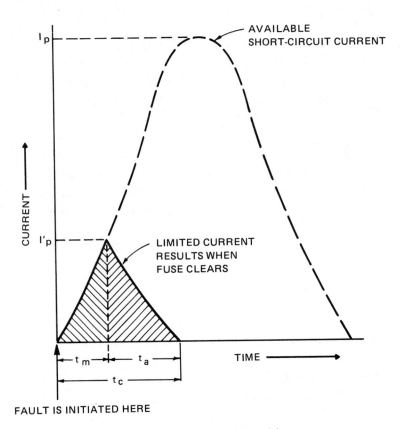

Fig. 18-4 Current-limiting effect of fuses.

All fuses carrying the Underwriters Laboratories Class listings (Class G, H, J, K, L, RK1, RK5, and T) and plug fuses are tested on alternating-current circuits and are marked for ac use. When fuses are to be used on direct-current systems, the electrician should consult the fuse manufacturer since it may be necessary to reduce the fuse voltage rating and the interrupting rating to insure safe operation.

The variables in the physical appearance of fuses include length, ferrule diameter, and blade length-width-thickness, as well as other distinctive features. For these reasons, it is difficult to insert a fuse of a given ampere rating into a fuseholder rated for less amperage than the fuse. The differences in fuse construction also make it difficult to insert a fuse of a given voltage into a fuseholder with a higher voltage rating, *NEC Section 240-60(b)*. Figure 18-5 indicates several examples of the method of insuring that fuses and fuseholders are not mismatched.

NEC Section 240-60(b) also specifies that fuseholders for current-limiting fuses shall be

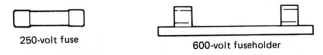

(A) A 250-volt fuse shall be constructed so that it cannot be inserted into a 600-volt fuseholder.

(B) A 60-ampere fuse is constructed so that it cannot be inserted into a 30-ampere fuseholder.

Fig. 18-5 Examples of *Section 240-60(b)* requirements.

designed so that they cannot accept fuses that are noncurrent limiting, figure 18-6.

In general, fuses may be used at system voltages that are less than the voltage rating of the fuse. For example, a 600-volt fuse combined with a 600-volt switch may be used on 575-volt, 480/277-volt, 208/120-volt, 240-volt, 50-volt, and 32-volt systems.

(A)

Fuse A is a Class H, noncurrent-limiting fuse. This fuse does not have the notch required to match the rejection pin in the fuse clip of the Class R fuseblock (C).

(B)

Fuse B is a Class R fuse (either Class RK1 or RK5), which is a current-limiting fuse. This fuse has the required notch on one blade to match the rejection pin in the fuse clip of the Class R fuseblock (C).

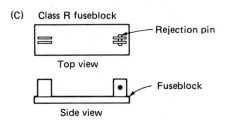

Fig. 18-6 Examples of *Section 240-60(b)* requirements.

Fuses that have silver links perform the best, followed by fuses with copper links, followed by fuses with zinc links.

Class H. Class H fuses, figure 18-7, formerly were called *NEC* or Code fuses. Most low-cost, common, standard nonrenewable one-time fuses are Class H fuses. Renewable-type fuses also come under the Class H classification. Neither the interrupting rating nor the notation *Class H* appears on the label of a class H fuse. This type of fuse is tested by the Underwriters Laboratories on circuits which deliver 10,000 amperes ac. Class H fuses are available with ratings ranging from 0-600 amperes in both 250-volt ac and 600-volt ac types. Class H fuses are not current limiting.

A higher quality nonrenewable one-time fuse is also available, called a Class K5 fuse, which has a 50,000-ampere interrupting rating. It is easy to identify this better grade of fuse because the *Class K5* and its interrupting rating are marked on the label.

Class K. Class K fuses are grouped into three categories: K1, K5, and K9, figure 18-8 A through D. These fuses may be UL listed with interrupting ratings in symmetrical rms amperes in values of 50,000, 100,000 or 200,000 amperes. For each K rating, UL has assigned a maximum level of peak let-through current (I_p) and energy as given by

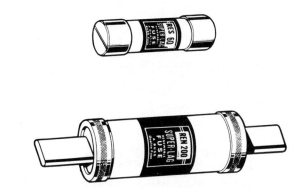

Fig. 18-7 Class H cartridge fuse. Illustration shows renewable-type fuse in which the blown link may be replaced.

$I^2 t$. Class K fuses have varying degrees of current-limiting ability, depending upon the K rating. Class K1 fuses have the greatest current-limiting ability and Class K9 fuses the least ability. Class K fuses may be listed as time-delay fuses as well. In this case, UL requires that the fuses have a minimum time delay of 10 seconds at 500% of the rated current. Class K fuses are available in ratings ranging from 0-600 amperes at 250- or 600-volts ac. Class K fuses have the same dimensions as Class H fuses.

Class J. Class J fuses are current limiting and are so marked, figure 18-9 A and B. They are listed by UL with an interrupting rating of 200,000 symmetrical rms amperes. Class J fuses are physically smaller than Class H fuses. Therefore, when a fuseholder is installed to accept a Class J fuse, it will be impossible to install a Class H fuse in the fuseholder, *NEC Section 240-60(b)*.

The Underwriters Laboratories has assigned maximum values of $I^2 t$ and I_p which are slightly less than those for Class K1 fuses. Both fast-acting, current-limiting Class J fuses and time-delay, current-limiting Class J fuses are available in ratings ranging from 0-600 amperes at 600-volts ac.

Class L. Class L fuses, figure 18-10 A and B, are listed by UL is sizes ranging from 601-6,000 amperes at 600 volts ac. These fuses have specified maximum values of $I^2 t$ and I_p. They are current-limiting fuses and have an interrupting rating of 200,000 symmetrical rms amperes. These bolt-type fuses are used in bolted pressure contact switches. Class L fuses are available in both a fast-acting, current-limiting type and a time-delay, current-limiting type. Both types of Class L fuses meet UL requirements.

Class T. Class T fuses, figure 18-11, are current-limiting fuses and are so marked. These fuses are UL listed with an interrupting capacity of 200,000 symmetrical rms amperes. Class T fuses are physically smaller than Class H or Class J fuses. The configuration of this type of fuse limits its use to fuseholders and switches that will reject all other types of fuses.

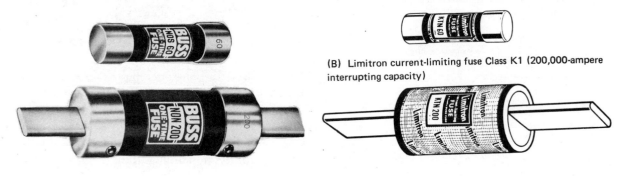

(A) One-time fuse Class K5 (50,000-ampere interrupting capacity)

(B) Limitron current-limiting fuse Class K1 (200,000-ampere interrupting capacity)

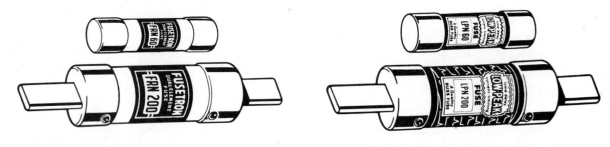

(C) Fusetron dual-element fuse Class K5 (200,000-ampere interrupting capacity)

(D) Low-peak dual-element current-limiting fuse, Class K1 (200,000-ampere interrupting capacity)

Fig. 18-8 Class K cartridge fuses.

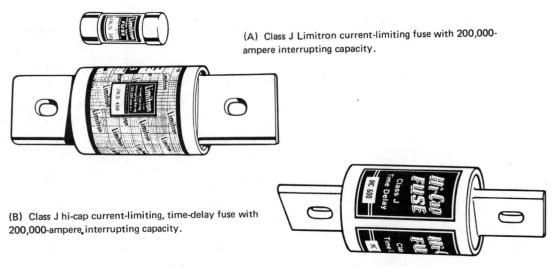

(A) Class J Limitron current-limiting fuse with 200,000-ampere interrupting capacity.

(B) Class J hi-cap current-limiting, time-delay fuse with 200,000-ampere interrupting capacity.

Fig. 18-9 Class J current-limiting fuses.

(A) Limitron current-limiting fuse, Class L, with 200,000-ampere interrupting capacity.

(B) Hi-cap current-limiting, time-delay fuse, Class L with 200,000-ampere interrupting capacity.

Fig. 18-10 Class L cartridge fuses.

Class T fuses have electrical characteristics similar to those of Class J fuses and are tested in a similar manner by UL. These fuses are available in ratings ranging from 0-600 amperes at both 300 volts and 600 volts. The 300-volt rating enables these fuses to be installed on 480/277 volts wye-connected electrical systems where the voltage line-to-ground is 277 volts, and the line-to-line voltage is 480 volts.

Class T fuses may be used to protect mains, feeders, branch circuits, and circuit breakers when the equipment is designed to accept this type of fuse.

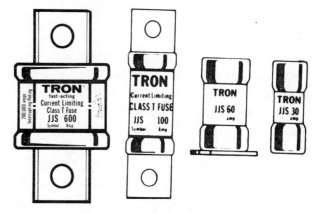

Fig. 18-11 Class T cartridge fuses.

Class G. Class G fuses, figure 18-12, are cartridge fuses with small physical dimensions. They are used on circuits of 300 volts or less to ground (for example, 277/480-volt systems). Class G fuses are available in sizes ranging from 0-60 amperes and are UL listed at an interrupting capacity of 100,000 symmetrical rms amperes. Commonly called Type SC fuses, they are essentially cartridge Type S fuses. To prevent overfusing, Class G fuses are size-limiting within the four categories assigned to their ampere ratings. Therefore, a fuseholder designed to accept a 15-ampere Type SC fuse will not accept a 20-ampere Type SC fuse; and a fuseholder designed to accept a 20-ampere Type SC fuse will not accept a 30-ampere Type SC fuse; and so on for the four categories.

Class R. The Class R fuse is another recent development in the UL standards listing for fuses. This

Ampere Rating	Dimensions
0–15	13/32″ X 1 5/16″
16–20	13/32″ X 1 13/32″
21–30	13/32″ X 1 5/8″
31–60	13/32″ X 2 1/4″

Fig. 18-12 Class G fuses.

fuse is a nonrenewable cartridge type and has an interrupting rating of 200,000 rms symmetrical amperes. The peak let-through current (I_p) and the total clearing energy (I^2t) values are specified for the individual case sizes. The values for I^2t and I_p are specified by UL based on short-circuit tests at 50,000, 100,000, and 200,000 amperes.

Class R fuses are divided into two subclasses: Class RK1 and Class RK5. The Class RK1 fuse has characteristics similar to the Class K1 fuse. The Class RK5 fuse has characteristics similar to the Class K5 fuse. These fuses must be marked either Class RK1 or RK5. In addition, they are marked to be current limiting.

The ferrule-type Class R fuse has a rating range of 0-60 amperes and can be distinguished by the annular ring on one end of the case, figure 18-13A. The knife blade-type Class R fuse has a rating range of 61-600 amperes and has a slot in the blade on one end, figure 18-13B. When a fuseholder is designed to accept a Class R fuse, it will be impossible to install a standard Class H or Class K fuse (these fuses do not have the annular ring or slot of the Class R fuse). The requirements for noninterchangeable cartridge fuses and fuseholders are covered in *NEC Section 240-60(b)*. However, the Class R fuse can be installed in older style fuse clips on existing installations. As a result, the Class R fuse may be called a *one-way rejection fuse*.

Electrical equipment manufacturers will provide the necessary rejection-type fuseholders in their equipment which is then tested with a Class R fuse at short-circuit current values of 50,000, 100,000, or 200,000 amperes. Each piece of equipment will be marked accordingly.

Plug Fuses (*NEC Article 240, Part E*)

Sections 240-50 through *240-54* of the *National Electrical Code* cover the requirements for plug fuses. The electrician will be most concerned with the following requirements for plug fuses, fuseholders, and adapters:

- they shall not be used in circuits exceeding 125 volts between conductors, except on systems having a grounded neutral with no conductor having more than 150 volts to ground.

This situation is found in the 120/208-volt system in the commercial building covered in this text, or in the case of a 120/240-volt, single-phase system.

- they shall have ampere ratings of 0 to 30 amperes.
- they shall have a hexagonal configuration for ratings of 15 amperes and below.
- the screw shell must be connected to the load side of the circuit.
- Edison-base plug fuses may be used only as replacements in existing installations where there is no evidence of overfusing or tampering.
- all new installations shall be in fuseholders requiring Type S plug fuses or fuseholders with a Type S adapter inserted to accept Type S fuses.
- Type S plug fuses are classified 0 to 15 amperes; 16 to 20 amperes; and 21 to 30 amperes.

Prior to the installation of fusible equipment, the electrician must determine the ampere rating of the various circuits. An adapter of the proper size is then inserted into the Edison-base fuseholder; finally, the proper size of Type S fuse can be inserted into the fuseholder. The adapter makes the fuseholder nontamperable and noninterchangeable. For example, after a 15-ampere adapter is inserted into a fuseholder for a 15-ampere circuit, it is impossible to insert a Type S fuse having a larger ampere rating into this fuseholder without removing the 15-ampere adapter. Adapters are designed so that they are extremely difficult to remove.

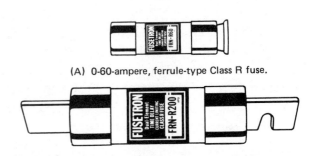

(A) 0-60-ampere, ferrule-type Class R fuse.

(B) 61-600-ampere, knife blade-type Class R fuse.

Fig. 18-13 Class R cartridge fuses (may be RK1 or RK5).

(A) Dual-element
type S fuse

(B) Adapter for
type S fuses

(C) Ordinary plug fuse, nontime-
delay, Edison-base type

(D) Fusetron dual-element,
time-delay plug fuse,
Edison-base type

Fig. 18-14 Type S fuses and adapter.

Type S fuses and suitable adapters, figure 18-14, are available in a large selection of ampere ratings ranging from 3/10 ampere to 30 amperes. When a Type S fuse has dual elements and a time-delay feature, it does not blow unnecessarily under momentary overloads, such as the current surge caused by the startup of an electric motor. On heavy overloads or short circuits, this type of fuse opens very rapidly.

Table 18-1 provides rating and application information for the fuses supplied by one manufacturer. The table indicates the Class, voltage rating, current range, interrupting ratings, and typical applications for the various fuses. A table of this type may be used to select the proper fuse to meet the needs of a given situation.

Cable Limiters

Cable limiters are used quite often in commercial and industrial installations where parallel cables are used on service entrances and feeders.

Cable limiters differ in purpose from fuses which are used for overload protection. Cable limiters are short-circuit devices that can *isolate* a faulted cable, rather than having the fault open

the entire phase. They are selected on the basis of conductor size; for example, a 500 MCM cable limiter would be used on a 500 MCM conductor.

Cable limiters are available for cable-to-cable or cable-to-bus installation, for either aluminum or copper conductors. The type of cable limiter illustrated in figure 18-15 is for use with a 500 MCM copper conductor, and is rated at 600 volts with an interrupting rating of 200,000 amperes.

Cable limiters are also used where taps are made ahead of the mains, such as they are in the commercial building where the emergency system is tapped ahead of the main fuses (as shown in figure 17-2).

Figure 18-16 is an example of how cable limiters may be installed on the service of the commercial building. A cable limiter is installed at each end of each 500 MCM conductor. Three conductors per phase terminate in the main switchboard where the service is then split into two 800-ampere bolted pressure switches.

Cable limiters are permitted by the *National Electrical Code* in *Section 230-82, Exception No. 1.*

Cable limiters are installed in the "hot" phase conductors only. They are not installed in the grounded neutral conductor.

Fig. 18-15 Cable limiter for 500 MCM copper conductor.

TABLE 18-1 BUSS FUSES
THEIR RATINGS, CLASS, AND APPLICATION**

FUSE & AMPERE RATING RANGE	Voltage Rating	Symbol	UL Class	Interrupting Rating in Sym. rms Amperes	Application All Types Recommended For Protection of General Purpose Circuits & Components
HI-CAP time-delay FUSE 601-6000A	600 V	KRP-C	L	200,000 A	High Capacity Main, Feeder & Br. Ckts. Large Motors Circuits. Has more time-delay than KTU.
LIMITRON fast-acting FUSE 601-6000A	600 V	KTU	L	200,000 A	High Capacity Main, Feeder, & Br. Ckts. Circuit Breaker Protection
FUSETRON dual-element FUSE 0-600A	250 V	FRN-R	RK5	200,000 A	Main, Feeder, & Br. Ckts. Especially Recommended For Motors Welders & Transformers
	600 V	FRS-R	RK5		
LOW-PEAK dual-element FUSE 0-600A	250 V	LPN-R	RK1	200,000 A	Main, Feeder & Br. Ckts. Especially Recommended For Motors Welders & Transformers. (More Current-limiting than Fusetron dual-element Fuse 0-600A.)
	600 V	LPS-R	RK1		
LIMITRON fast-acting FUSE 0-600A	250 V	KTN-R	RK1	200,000 A	Main, Feeder & Br. Ckts. Especially Recommended For Circuit Breaker Protection. (High Degree of Current-limitation)
	600 V	KTS-R	RK1		
TRON fast-acting FUSE 0-600A	250 V 600 V	JJN JJS	T	200,000 A	Main, Feeder & Br. Ckts. Circuit Breaker Protection Small Physical Dimensions (High Degree of Current-Limitation)
HI-CAP time-delay FUSE 15-600A	600 V	JHC	*	200,000 A	Main, Feeder & Br. Ckts., Motor & Transformer Ckts.
LIMITRON quick-acting FUSE 0-600A	600 V	JKS	J	200,000 A	Main, Feeder & Br. Ckts. Circuit Breaker Protection
ONE-TIME FUSE 0-600A	250 V 600 V	NON NOS	H	10,000 A	General Purpose
SC FUSE 0-60A	300 V	SC	G	100,000 A	General Purpose Branch Circuits.
FUSETRON dual-element PLUG FUSE 0-30A, Edison Base	125 V	T	⚱	10,000 A	General Purpose (Ideal for Motors)
FUSTAT dual-element type S PLUG FUSE 0-30A	125 V	S	S	10,000 A	General Purpose (Ideal for Motors)
ORDINARY PLUG FUSE Edison Base, 0-30A	125 V	W	⚱	10,000 A	General Purpose Less Time Delay than Types T & S

* JHC Fuse Has Class J Dimensions but With More Time-Delay in Electrical Characteristics. Listed by CSA to their maximum test std. of 100,000 A.

** For more finite applications consult electrical equipment manufacturer and fuse bulletin.

⚱ UL Listed as Edison Base Plug Fuse.

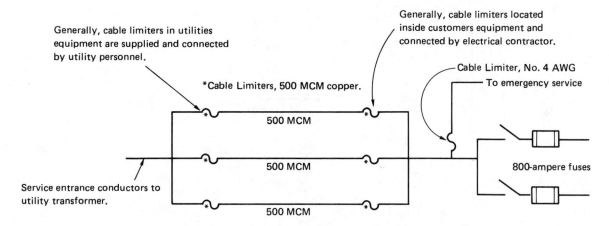

Generally, cable limiters in utilities equipment are supplied and connected by utility personnel.

Generally, cable limiters located inside customers equipment and connected by electrical contractor.

Cable Limiter, No. 4 AWG
To emergency service

*Cable Limiters, 500 MCM copper.

500 MCM

500 MCM

800-ampere fuses

Service entrance conductors to utility transformer.

500 MCM

Fig. 18-16 Use of a cable limiter in the service of the commercial building.

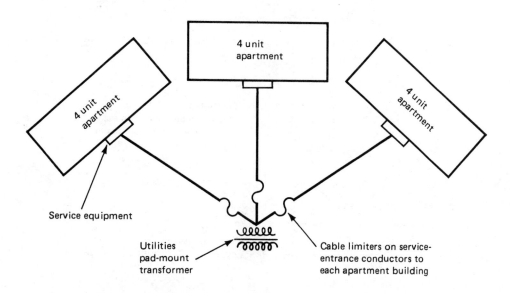

4 unit apartment

4 unit apartment

4 unit apartment

Service equipment

Utilities pad-mount transformer

Cable limiters on service-entrance conductors to each apartment building

Fig. 18-17 Some utilities install cable limiters where more than one customer is connected to one transformer. Thus, should one customer have problems in their main service equipment, the cable limiter on that service will open, isolating the problem from other customers on the same transformer. This is common for multiunit apartments, condos and shopping centers.

DELTA, THREE-PHASE, GROUNDED "B" PHASE SYSTEM

Fuses shall be installed in series with un-grounded conductors for overcurrent protection, *Section 240-20*. The Code prohibits overcurrent devices from being connected in series with any conductor that is intentionally grounded, *Section 240-22*.

However, there are certain instances where a three-pole, three-phase switch may be installed, where it is permitted to install fuses in two poles only. This would be in the case of a delta, three-phase, corner grounded "B" phase system, figure 18-18. A device called a *solid neutral* can be installed in the "B" phase, as shown in the figure. Note that the switch has two fuses and one solid neutral installed.

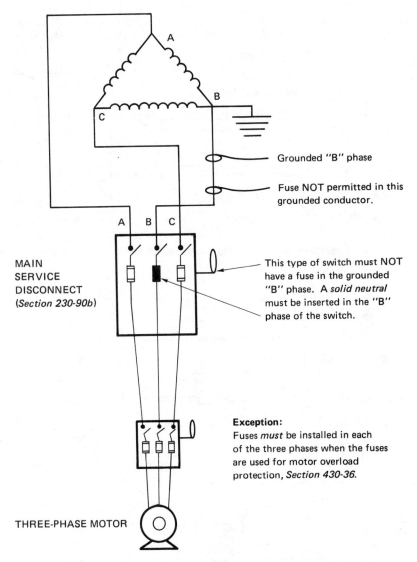

Grounded "B" phase

Fuse NOT permitted in this grounded conductor.

MAIN
SERVICE
DISCONNECT
(*Section 230-90b*)

This type of switch must NOT have a fuse in the grounded "B" phase. A *solid neutral* must be inserted in the "B" phase of the switch.

Exception:
Fuses *must* be installed in each of the three phases when the fuses are used for motor overload protection, *Section 430-36.*

THREE-PHASE MOTOR

Fig. 18-18 Three-phase, three-wire delta system with grounded "B" phase.

Solid Neutrals

A solid neutral, figure 18-19, is made of copper bar that has exactly the same dimensions as a fuse for a given ampere rating and voltage rating. For example, a solid neutral rated at 100 amperes, 600 volts would be installed in a switch rated at 100 amperes, 600 volts.

TIME-CURRENT CHARACTERISTIC CURVES AND PEAK LET-THROUGH CHARTS

The electrician must have a basic amount of information concerning fuses and their application to be able to make the correct choices and decisions for everyday situations that arise on an installation.

The electrician must be able to use the following types of fuse data:

- time-current characteristic curves, including total clearing and minimum melting curves.

- peak let-through charts.

Fuse manufacturers furnish this information for each of the fuse types they produce.

The Use of Time-current Characteristic Curves

The use of the time-current characteristic curves shown in figures 18-20 and 18-21 can be demonstrated by considering a typical problem. Assume that an electrician must select a fuse to protect a motor that is governed by *Section 430-*

Fig. 18-19 Examples of a solid neutral.

32(a). This section indicates that the running over-current protection shall be based on not over 125% of the full-load running current of the motor. The motor in this example is assumed to have a full-load current of 24 amperes. Therefore, the size of the required protective fuse is determined as follows:

$$24 \times 1.25 = 30 \text{ amperes}$$

Now the question must be asked: is this 30-ampere fuse capable of holding the inrush current of the motor (which is approximately four to five times the running current) for a sufficient length of time for the motor to reach its normal speed? For this problem, the inrush current of the motor is assumed to be 100 amperes.

Refer to figure 18-20. At the 100-ampere line, draw a horizontal line until it intersects with the 30-ampere fuse line. Then draw a vertical line down to the base line of the chart. At this point, it can be seen that the 30-ampere fuse will hold 100 amperes for approximately 40 seconds. In this amount of time, the motor can be started. In addition, the fuse will provide running overcurrent protection as required by *NEC Section 430-32(a)*.

If the time-current curve for thermal overload relays in a motor controller is checked, it will show that the overload element will open the same current in much less than 40 seconds. Therefore, in the event of an overload, the thermal overload elements will operate before the fuse opens. If the overload elements do not open for any reason, or if the contacts of the controller weld together, then the properly sized dual-element fuse will open to reduce the possibility of motor burnout. The above method is a simple way to obtain back-up or double motor protection.

Referring again to figure 18-20, assume that a short circuit of approximately 500 amperes

occurs. Find the 500-ampere line at the left of the chart and then, as before, draw a horizontal line until it intersects with the 30-ampere fuse line. From this intersection point, drop a vertical line to the base line of the chart. This value indicates that the fuse will open the fault in slightly over 2/100 (0.02) seconds. On a 60-hertz system (60 cycles per second), one cycle equals 0.016 seconds. Therefore, a 30-ampere fuse will clear a 500-ampere fault in just over one cycle.

The Use of Peak Let-through Charts

The withstand ratings of electrical equipment (such as bus duct, switchboard bracing, controllers, and conductors) and the interrupting ratings of circuit breakers are given in the published standards of the Underwriters Laboratories, NEMA, and the Insulated Power Cable Engineering Association (IPCEA). The withstand ratings may be based either on the peak current (I_p), or on the root mean square (rms) current.

As an example of the use of the let-through chart in figure 18-22, assume that it is necessary to protect molded case circuit breakers. These breakers have an interrupting rating of 10,000 amperes rms symmetrical. The available short-circuit current at the panel is 40,000 amperes rms symmetrical. The fuse protecting the circuit breaker panel is rated at 100 amperes. This 100-ampere fuse may be located either in the panel or at some distance away from the panel in a separate disconnect switch placed at the main service or distribution panel, figure 18-23.

Now refer to figure 18-22. Find the 40,000-ampere point on the base line of the chart and then draw a vertical line upward until it intersects the 100-ampere fuse curve. Move horizontally to the left until Line A-B is reached. Then move vertically downward to the base line. The value of this point on the base line is 4,600 amperes and is the apparent rms amperes. This means that the current-limiting effect of the fuse, when subjected to a 40,000-ampere rms symmetrical fault, permits an apparent let-through current of only 4,600 amperes rms symmetrical. Therefore, the 100-ampere fuse selected readily protects the circuit breaker (with a 10,000-ampere interrupting rating) against the 40,000-

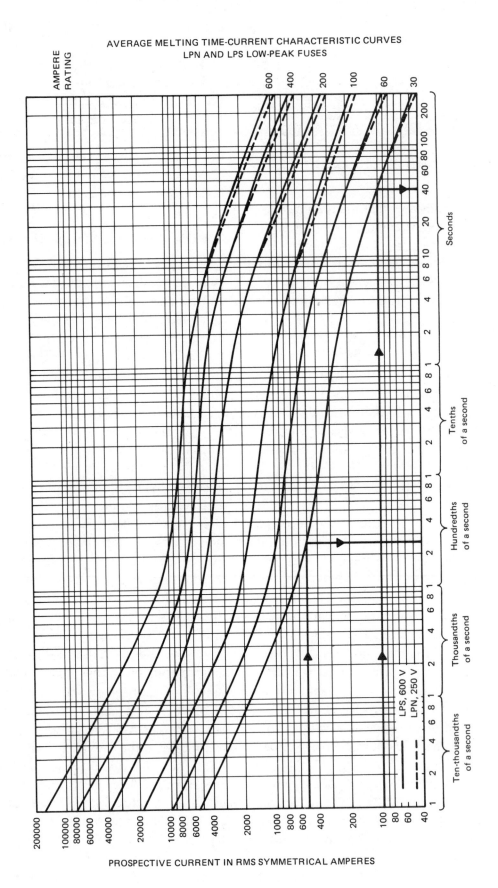

Fig. 18-20

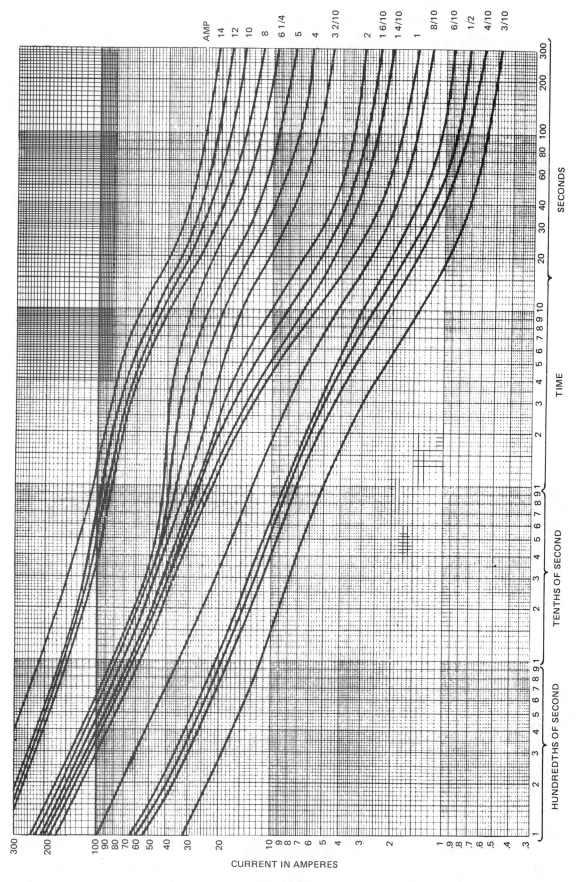

Fig. 18-21

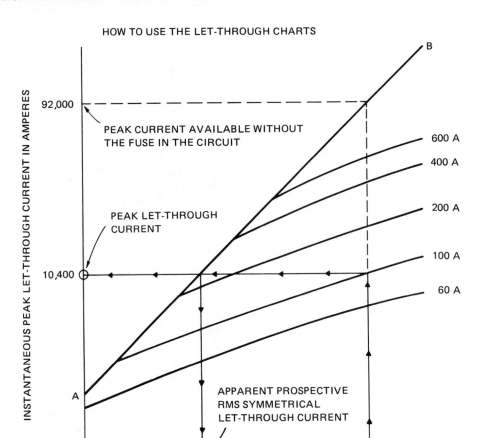

Fig. 18-22 Using the let-through charts to determine peak let-through current and apparent prospective rms symmetrical let-through current.

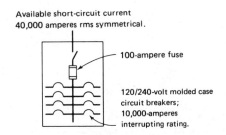

Fig. 18-23 Example of a fuse used to protect circuit breakers. This installation meets the *National Electrical Code* requirements outlined in *Sections 110-9* and *230-98*.

ampere rms symmetrical fault. Thus, the requirements of *NEC Sections 110-9* and *230-98* are met. Actual tests conducted by manufacturers have shown that the same fuse can be used to protect a circuit breaker with a 10,000-ampere interrupting rating when 100,000 amperes are available.

In figure 18-22, note that at a current of 40,000 amperes rms symmetrical, the peak current,

I_p, available at the first half cycle is 92,000 amperes if there is no fuse in the circuit to limit the let-through current. When the 100-ampere fuse is installed, the instantaneous peak let-through current, I_p, is limited to just 10,400 amperes.

The magnetic stresses which occur in the example given can be compared. When the current-limiting fuse is installed, the magnetic stresses are only 1/80 of the stresses present when a noncurrent-limiting overcurrent device is used to protect the circuit.

Figure 18-24 shows a peak let-through chart for a family of dual-element fuses. Fuse manufacturers provide peak let-through charts for the various sizes and types of fuses they produce.

Tables A and B in figure 18-25 are typical of the tables published by circuit breaker manufacturers to show the maximum fuse size that can be used to protect specific circuit breaker sizes.

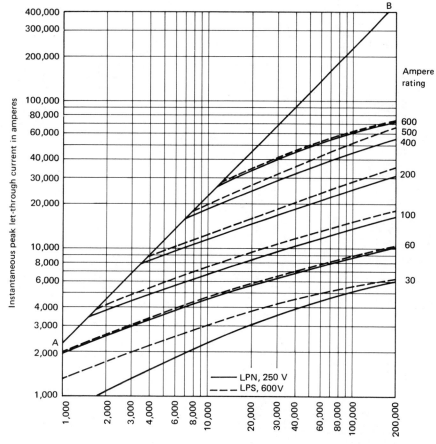

Fig. 18-24 Current-limiting effect LPN and LPS low-peak fuses.

CIRCUIT BREAKERS (*NEC Article 240, Part G*)

The *NEC* defines a circuit breaker as a device which is designed to open and close a circuit by non-automatic means and to open the circuit automatically on a predetermined overcurrent without being damaged itself when properly applied within its rating.

Molded case circuit breakers are the most common type in use today, figure 18-26. The tripping mechanism of this type of breaker is enclosed in a molded plastic case. The *thermal-magnetic* type of circuit breaker is covered in this unit. Another type is the power circuit breaker which is larger, of heavier construction, and more costly. Power circuit breakers are used in large industrial applications. However, these breakers perform the same electrical functions as the smaller molded case breakers.

Circuit breakers are also available with solid-state, adjustable trip ratings and settings. When compared to the standard molded case circuit breaker, these breakers are very costly.

Molded case circuit breakers are listed by the Underwriters Laboratories and are covered by NEMA Standards. *NEC Sections 240-80* through *240-83* state the basic requirements for circuit breakers and are summarized as follows.

- Breakers shall be trip free so that if the handle is held in the *On* position, the internal mechanism trips the breaker to the *Off* position.

- The breaker shall clearly indicate if it is in the *On* or *Off* position.

- The breaker shall be nontamperable; that is, it cannot be readjusted (to change its trip point or time required for operation) without dismantling the breaker or breaking the seal.

- The rating shall be durably marked on the breaker. For the smaller breakers with ratings

TABLE A — STANDARD CIRCUIT BREAKERS*

Frame	Ampere Rating	Minimum Fuse Rating Circuit Breaker Inst. Set		Maximum Fuse Rating			
				600 & 480 Volts A-C		240 Volts A-C	
		Lo	High	Load Side	Line Side	Load Side	Line Side
50 A (5000A IC)	15–20	—	50	—	—	—	200
	30–50	—	100	—	—	—	200
100A (7500– 15000A IC)	15	—	70	200	300	400	400
	20–40	—	100	200	300	400	400
	50–100	—	200	200	300	400	400
225A (10000A IC)	125–225	—	400	—	—	400	400
225 A (15000– 25000A IC)	70–150	300	400	600	600	800	800
	175	300	400	600	1000	1000	1000
	200–225	300	600	600	1000	1000	1000
400A (25000– 50000A IC)	70–125	300	400	800	1000	1000	1000
	150–175	300	400	800	1000	1000	1000
	200–225	300	600	800	1000	1000	1000
	250–300	400	600	800	1200	1200	1200
	350–400	600	800	800	1200	1200	1200
800A (25000– 50000A IC)	125–175	300	400	800	1000	1000	1000
	200–225	300	600	800	1000	1000	1000
	250–275	400	600	800	1000	1000	1000
	300	400	600	800	1200	1200	1200
	350	600	800	800	1200	1200	1200
	400	600	800	1200	1200	1200	1200
	500–600	800	1200	1200	1200	1200	1200
	700–800	1000	1200	1200	1200	1200	1200

TABLE B — HIGH INTERRUPTING CAPACITY CIRCUIT BREAKERS*

Frame	Ampere Rating	Minimum Fuse Rating Circuit Breaker Inst. Set.		Maximum Fuse Rating			
				600 & 480 Volts A-C		240 Volts A-C	
		Lo	High	Load Side	Line Side	Load Side	Line Side
100A (20000– 75000A IC)	15–20	—	70	400	600	1000	1000
	30–40	—	100	400	600	1000	1000
	50–70	—	200	600	800	1200	1200
	90–100	—	300	600	800	1200	1200
400A (30000– 75000A IC)	125	300	400	1200	1200	2000	2000
	150–175	300	400	1600	1600	2500	2500
	200–225	300	600	1600	2000	2500	2500
	250–300	400	600	1600	2000	2500	2500
	350–400	600	800	1600	2000	2500	2500
800A (30000– 75000A IC)	400	600	800	3000	3000	3000	3000
	500–600	800	1200	3000	3000	3000	3000
	700–800	1000	1200	3000	3000	3000	3000
1200 or 1600A (50000– 75000-IC)	800	1000	1200	3000	3000	3000	3000
	1000–1200	1600	2000	3000	3000	3000	3000
	1400–1600	2000	3000	3000	3000	3000	3000

NOTE: Tables A & B are applicable for Limitron and Hi-Cap Fuses on systems up to 100,000 RMS symmetrical amperes available.
    *See breaker manufacturer for more specific details.

Fig. 18-25 Typical tables published by circuit breaker manufacturers showing circuit breaker protection requirements.

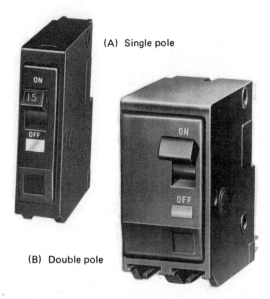

(A) Single pole

(B) Double pole

Fig. 18-26 Molded case circuit breakers.

Thermal-magnetic Circuit Breakers

A thermal-magnetic circuit breaker contains a bimetallic element. On a continuous overload, the bimetallic element moves until it unlatches the inner tripping mechanism of the breaker. Harmless momentary overloads do not cause the tripping of the bimetallic element. If the overload is heavy, or if a short circuit occurs, then the mechanism within the circuit breaker causes the breaker to interrupt the circuit instantly. The time required for the breaker to open the circuit completely depends upon the magnitude of the fault current and the mechanical condition of the circuit breaker. This time may range from approximately one-half cycle to several cycles.

of 100 amperes or less and 600 volts or less, this marking must be molded, stamped, or etched on the handle or other portion of the breaker that will be visible after the cover of the panel is installed.

- Every breaker having an interrupting rating other than 5,000 amperes shall have its interrupting rating shown on the breaker.

- The *NEC, Section 240-83(d),* and the Underwriters Laboratories require that when a circuit breaker is to be used as a toggle switch, the breaker must be so tested by the Laboratories. A typical use of a circuit breaker as a toggle switch is in a panel when it is desired to control lighting circuits (turn them on and off) instead of installing separate toggle switches. Breakers listed for use as toggle switches will bear the letters *SWD* on the label. Breakers not marked in this manner must not be used as switches.

- A circuit breaker should not be loaded to more than 80% of its current rating for loads which are likely to be on for three hours or more in a 24-hour period, unless the breaker is marked otherwise, such as those breakers tested by the Underwriters Laboratories at 100% loading.

Circuit breaker manufacturers calibrate and set the tripping characteristic for most molded case breakers. Breakers are designed so that it is difficult to alter the set tripping point, *Section 240-82.* For certain types of breakers, however, the trip coil can be changed physically to a different rating. Adjustment provisions are made on some breakers to permit the magnetic trip range to be changed. For example, a breaker rated at 100 amperes may have an external adjustment screw with positions marked HI-MED-LO. The manufacturer's application data for this breaker indicates that the magnetic tripping occurs at 1,500 amperes, 1,000 amperes, or 500 amperes respectively for the indicated positions. These settings usually have a tolerance of ±10%.

The *ambient-compensated* type of circuit breaker is designed so that its tripping point is not affected by an increase in the surrounding temperature. An ambient-compensated breaker has two elements: the first element heats due to the current passing through it and because of the heat of the surrounding air; the second element is affected only by the ambient temperature. These elements act in opposition to each other. In other words, as the tripping element tends to lower its tripping point because of a high ambient temperature, the second element exerts an opposing force which stabilizes the tripping point. Therefore, current through the tripping element is the only factor that causes the element to open the circuit.

Factors which can affect the proper operation of a circuit breaker include moisture, dust, vibration, corrosive fumes and vapors, and excessive tripping and switching. As a result, care must be taken when locating and installing circuit breakers and all other electrical equipment.

The interrupting rating of a circuit breaker is marked on the breaker label. The electrician should check the breaker carefully for the interrupting rating since the breaker may have several voltage ratings with a different interrupting rating for each. For example, assume that it is necessary to select a breaker having an interrupting rating at 240 volts of at least 50,000 amperes. A close inspection of the breaker may reveal the following data:

Voltage	Interrupting Rating
240 volts	65,000 amperes
480 volts	25,000 amperes
600 volts	18,000 amperes

(Recall that for a fuse, the interrupting rating is marked on the fuse label. This rating is the same for any voltage up to and including the maximum voltage rating of the fuse.)

The standard full-load ampere ratings of nonadjustable circuit breakers are the same as those for fuses according to *NEC Section 240-6*. As previously mentioned, additional standard ratings of fuses are 1, 3, 6, 10, and 601 amperes.

The time-current characteristics curves for circuit breakers are similar to those for fuses. A typical circuit breaker time-current curve is shown in figure 18-27. Note that the current is indicated in percentage values of the breaker trip unit rating. Therefore, according to this graph, the 100-ampere breaker being considered will:

1. carry its 100-ampere (100%) rating indefinitely.
2. carry 300 amperes (300%) for a minimum of 25 seconds and a maximum of 70 seconds.
3. unlatch its tripping mechanism in 0.0032 seconds (approximately one-quarter cycle) with a current of 5,000 amperes.
4. interrupt the circuit in a maximum time of 0.016 seconds (one cycle) with a current of 5,000 amperes (5,000%).

The same time-current curve can be used to determine that a 200-ampere circuit breaker will:

1. carry its 200-ampere (100%) rating indefinitely.
2. carry 600 amperes (300%) for a minimum of 25 seconds and a maximum of 70 seconds.
3. unlatch its tripping mechanism in 0.0032 seconds (approximately one-quarter cycle) with a current of 5,000 amperes.
4. interrupt the circuit in a maximum time of 0.016 seconds (one cycle) with a current of 5,000 amperes (2,500%).

This example shows that if a short circuit occurs in the order of 5,000 amperes, then both the 100-ampere breaker and the 200-ampere breaker installed in the same circuit will open together because they have the same unlatching times. In many instances, this action is the reason for otherwise unexplainable power outages (Unit 19).

Common Misapplication

A common violation of *NEC Sections 110-9, 110-10*, and *230-98* is the installation of a main circuit breaker (such as a 100-ampere breaker) which has a high interrupting rating (such as 50,000 amperes) while making the assumption that the branch-circuit breakers (with interrupting ratings of 5,000 amperes) are protected adequately against the 40,000-ampere short circuit, figure 18-28.

Circuit breaker manufacturers state that a breaker having a high interrupting rating cannot protect a circuit breaker having a lower rating. In addition, the manufacturers state that the rating of a panel is based on the interrupting rating of the lowest rated device in that panel.

Since the opening time of a 100-ampere circuit breaker generally exceeds one-half cycle, almost the full amount of current (40,000 amperes) will enter the branch-circuit breaker section of the panel where the breaker interrupting rating is 5,000 amperes. **The probable outcome of such a situation is an electrical explosion, serious personal injury, or property loss resulting from a fire.**

A similar situation exists when a fuse with an interrupting rating of 50,000 amperes is installed ahead of a plug or cartridge fuse having an interrupting rating of 10,000 amperes. As in the case of

Current in percent of breaker trip unit rating

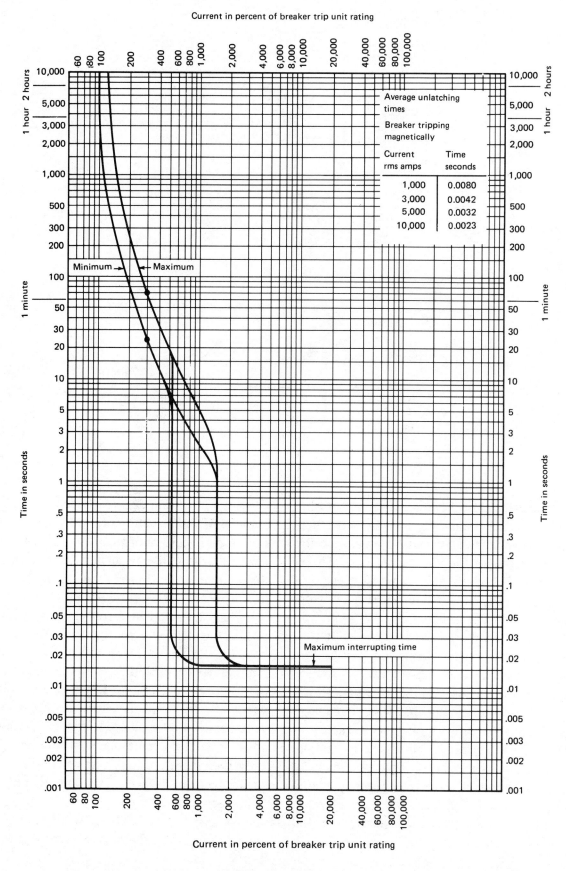

Fig. 18-27 Typical molded case circuit-breaker time-current curve.

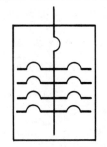

AVAILABLE SHORT-CIRCUIT CURRENT
AT PANEL 40,000 AMPERES.

*100-AMPERE MAIN BREAKER WITH INTER-
RUPTING RATING OF 50,000 AMPERES.

MOLDED CASE CIRCUIT BREAKERS; INTER-
RUPTING RATING OF 5,000 AMPERES.

*NOTE THAT THE MAIN BREAKER CANNOT PROTECT THE BRANCH
BREAKERS AGAINST A SHORT CIRCUIT OF 40,000 AMPERES; BRANCH
BREAKERS ARE CAPABLE OF INTERRUPTING 5,000 AMPERES.

Fig. 18-28 Common Code violation. *Sections 110-9* and *230-98.*

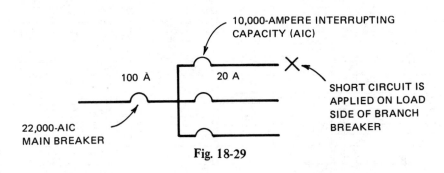

10,000-AMPERE INTERRUPTING
CAPACITY (AIC)

100 A   20 A   X

SHORT CIRCUIT IS
APPLIED ON LOAD
SIDE OF BRANCH
BREAKER

22,000-AIC
MAIN BREAKER

Fig. 18-29

circuit breakers, a fuse having a high interrupting rating cannot protect a fuse having a low interrupting rating against fault currents in excess of the interrupting rating of the lower rated fuse. Therefore, *all* fuses in a given panel must be capable of interrupting the maximum fault current available at that panel.

Underwriters Laboratories has tested a limited number of circuit-breaker panels and load centers under the test conditions of two breakers installed in series. In the example shown in figure 18-29, both the 20-ampere and the 100-ampere breakers will trip if they are properly matched. The series arrangement of breakers results in a cushioning effect when the two breakers trip simultaneously (this effect is sometimes called *cascading*). A panel tested under these conditions may be rated as an entire unit. For the situation in figure 18-29, the panel is rated as suitable for connection to a

system which can deliver no more than 22,000 amperes of short-circuit current.

A disadvantage of this type of panel is that for heavy short circuits on any of the branch circuits, the main breaker also trips and causes the complete loss of power to the entire panel.

Peak let-through charts are not available for circuit breakers due to the fact that the circuit breaker alone is not considered to be a current-limiting device. In other words, since the breakers cannot open a circuit in less than one-half cycle, nearly the full available short-circuit current must pass through the breaker. Therefore, it is important when installing circuit breakers to insure that not only the circuit breakers have the proper interrupting rating (capacity), but also that all of the components connected to the circuit downstream from the breakers are capable of withstanding the let-through current of the breaker. To use the previous problem as an example, this means that the branch-circuit conductors must be capable of withstanding 40,000 amperes for approximately one cycle (the opening time of the breaker).

Molded case circuit breakers with an integral current-limiting fuse are available. The thermal element in this type of breaker is used for low overloads, the magnetic element is used for low-level short circuits, and the integral fuse is used for short circuits of high magnitude. The interrupting capacity of this breaker/fuse combination (sometimes called a *limiter*) is higher than that of the same breaker without the fuse.

The selection of the ampacity of circuit breakers for branch circuits is governed by the requirements of *Article 240* of the *National Electrical Code.*

Table 430-152 shows that on the motor branch circuits, the maximum setting of an inverse-time circuit breaker should be sized at not over 250% of the full-load current of the motor. This table also shows that instantaneous trip breakers should be set at not over 700% of the motor full-load current.

NEC *Section 430-52(c)* makes an exception to the 700% rule and allows a maximum setting which is not to exceed 1,300% of the full-load current of the motor when the 700% setting is not sufficient for the starting current of the motor.

IMPORTANT: Carefully read the label on the equipment to be connected. When the label states "FUSES," fuses must be used for the overcurrent protection. Circuit breakers are not permitted in this case.

Some labels might read "maximum size overcurrent device" in which case fuses or circuit breakers could be used for branch-circuit overcurrent protection.

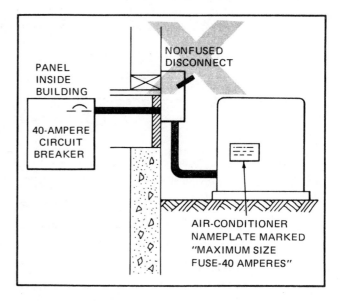

Fig. 18-31 This installation *violates* the Code, *Section 110-3(b).* Although the disconnect is within sight of the air conditioner, it does not contain fuses. Note that the branch-circuit protection is provided by the 40-ampere circuit breaker inside the building. Note also that the nameplate requires that the branch-circuit protection be 40-ampere fuses maximum. If fused branch-circuit protection were provided at the panel inside the building, the installation would meet Code requirements. Reprinted from Ray C. Mullin, *Electrical Wiring – Residential,* figure 23-11. © 1984 by Delmar Publishers Inc.

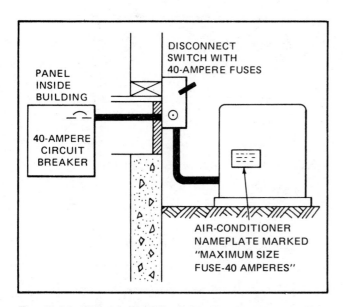

Fig. 18-30 This installation *conforms* to the Code, *Section 440-14.* The disconnect switch is within sight of the unit and contains the 40-ampere fuses called for on the air-conditioner nameplate as the branch-circuit protection. Reprinted from Ray C. Mullin, *Electrical Wiring – Residential,* figure 23-10. © 1984 by Delmar Publishers Inc.

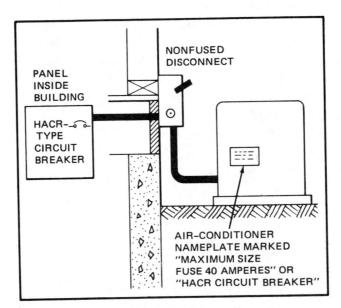

Fig. 18-32 Type HACR circuit breakers (two-pole, 15 through 60 amperes, 120/240 volt) are UL listed for group motor application such as found on heating (H), air-conditioning (AC), and refrigeration (R). The label on these breakers will have the letters HACR. The equipment must also be tested and marked, "Suitable for protection by a Type HACR breaker." If the HACR marking is not found on both the circuit breaker and the equipment, the installation does not meet the Code, unless fuses are installed for the branch-circuit overcurrent protection. Reprinted from Ray C. Mullin, *Electrical Wiring – Residential,* figure 23-12. © 1984 by Delmar Publishers Inc.

REVIEW

Note: Refer to the *National Electrical Code* or the plans as necessary.

1. What is the purpose of overcurrent protection? _____

    _____

    _____

    _____

2. List the four factors that must be considered when selecting overcurrent protective devices.

    1) _____ ,   3) _____ ,

    2) _____ ,   4) _____ .

3. What sections of the *National Electrical Code* require that overcurrent protective devices have adequate interrupting ratings? _____

    _____

4. Indicate the ampere rating of the fuse or circuit breaker that is selected to protect the following copper conductors. (Refer to *NEC Sections 240-3* and *240-6,* and *Table 310-16* and *footnote.*)

    a. No 12 TW _____-ampere overcurrent device.

    b. No. 8 TW _____-ampere overcurrent device

    c. No. 3 THW _____-ampere overcurrent device

    d. No. 3/0 RHW _____-ampere overcurrent device

5. A motor draws a full-load current of 40 amperes. Complete parts (a), (b), and (c) for this motor.
   a. Install _____ -ampere dual-element fuses in a _____ -ampere switch.
   b. Install _____ -ampere current-limiting fuses in a _____ -ampere switch.
   c. Install a _____ -ampere circuit breaker.

6. Current-limiting fuses generally contain _____ links surrounded by _____ arc-quenching filler.

7. In your own words, explain the meaning of the phrase *interrupting rating*.

   _____

   _____

   _____

   _____

   _____

8. In your own words, explain the meaning of the phrase *current limiting*.

   _____

   _____

   _____

   _____

   _____

9. Define the following terms:
   a. $I^2 t$ _____
   b. I_p _____
   c. rms _____

10. Provide the correct information for the fuse classes indicated:

	Ampere Range	Voltage Rating	Interrupting Rating
Class G fuses	_____ A	_____ V	_____ A
Class H fuses	_____ A	_____ V	_____ A
Class K fuses	_____ A	_____ V	_____ A
Class J fuses	_____ A	_____ V	_____ A
Class L fuses	_____ A	_____ V	_____ A
Plug fuses	_____ A	_____ V	_____ A
Class R fuses	_____ A	_____ V	_____ A
Class T fuses	_____ A	_____ V	_____ A

11. Class R, Class T, and Class J fuses (will) (will not) fit into standard fuse clips. (Underline the correct answer.)

12. Using figure 18-20, determine the opening time for a 60-ampere fuse which is loaded to 300 amperes. _____ second(s)

13. A short-circuit current of 800 amperes will open a 60-ampere fuse of the type shown in figure 18-20 in _____ second(s). A short-circuit current of 5,000 amperes will open the same fuse in _____ second(s).

14. Using the chart in figure 18-24, determine the approximate instantaneous peak let-through values and the apparent (equivalent) values of current for the following short-circuit currents.

Prospective Short-circuit Current	Fuse	Instantaneous Peak Let-through Amperes	Apparent rms Current
a. 100,000 A	60 A, 250 V	_____ A	_____ A
b. 50,000 A	60 A, 250 V	_____ A	_____ A
c. 200,000 A	200 A, 600 V	_____ A	_____ A

15. A section of plug-in bus duct is braced for 14,000 amperes rms. The available short-circuit current is 30,000 amperes rms symmetrical. Using figure 18-24, determine if a 200-ampere, 250-volt, low-peak, dual-element fuse will limit the current sufficiently to protect the bus duct against a short circuit. Explain. _____

_____

_____

_____

16. If the same section of plug-in bus duct (question 15) is connected to a 200-ampere standard molded case circuit breaker having an interrupting rating of 42,000 amperes, will the duct be protected properly? Explain. _____

_____

_____

_____

17. In a thermal-magnetic circuit breaker, overloads are sensed by the _____ element; short circuits are sensed by the _____ element.

18. Is it possible to install:

a. a 20-A, Type S fuse in a 15-A adapter? _____

b. a 30-A, Type SC fuse in a 20-A fuseholder? _____

c. a 100-A, Class H fuse in a 100-A fuseholder for Class J fuses? _____

d. a Class L fuse in a 400-A standard switch? _____

19. A cable limiter

a. is a short-circuit device only.   T   F

b. is not to be used for overload protection.   T   F

c. is generally connected to both ends of large paralleled conductors so that if a fault occurs on one of the conductors, that faulted cable is isolated from the system.   T   F

d. is rated at 600 V, 200,000 A interrupting capacity.   T   F

20. When the label on equipment states "Maximum size fuse 50 amperes" is it permitted to connect the equipment to a 50-ampere circuit breaker?

UNIT 19

Short-circuit Calculations and Coordination of Overcurrent Protective Devices

OBJECTIVES

After completing the study of this unit, the student will be able to

- perform short-circuit calculations using the point-to-point method.
- calculate short-circuit currents using the appropriate tables and charts.
- define the terms coordination, selective systems, and nonselective systems.
- define the term interrupting rating and explain its significance.
- use time-current curves.

The student must understand the intent of *National Electrical Code Sections 110-9, 110-10, and 230-98.* That is, to insure that the fuses and/or circuit breakers selected for an installation are capable of interrupting the current at the rated voltage that may flow under any condition (overload, short circuit, or ground fault) with complete safety to personnel and without damage to the panel, load center, switch, or electrical equipment in which the protective devices are installed.

An overloaded condition resulting from a miscalculation of load currents will cause a fuse to blow or a circuit breaker to trip in a normal manner. However, a miscalculation, a guess, or ignorance of the magnitude of the available short-circuit currents may result in the installation of breakers or fuses having inadequate interrupting ratings. Such a situation can occur even though the load currents in the circuit are checked carefully. **Breakers or fuses having inadequate interrupting ratings need only be subjected to a short circuit to cause them to explode, resulting in injury to personnel and serious damage to the electrical equipment.** *The interrupting rating of an overcurrent device is its maximum rating and must not be exceeded.*

In any electrical installation, the size of a given panel is determined first; for example, the installation may require 20 circuits at 20 amperes each. The next step is to determine the interrupting rating requirements of the fuses or circuit breakers to be installed in the panel. *NEC Section 110-9* is an all-encompassing section that covers the interrupting rating requirements for services, mains, feeders, subfeeders, and branch-circuit overcurrent of devices. *NEC Section 230-98* emphasizes adequate interrupting ratings specifically for services. For various types of equipment, normal currents can be determined by checking the equipment nameplate current, voltage, and wattage ratings. In addition, an ammeter can be used to check for normal and overloaded circuit conditions.

A number of formulas can be used to calculate short-circuit currents. Manufacturers of fuses, circuit breakers, and transformers publish numerous tables and charts showing approximate values of short-circuit current. This type of information is also provided by many NEMA standards. A standard ammeter must *not* be used to read short-circuit current, as this practice will result in damage to the ammeter and possible injury to personnel.

SHORT-CIRCUIT CALCULATIONS

The following sections will cover several of the basic methods of determining available short-circuit currents. As the short-circuit values given in the various tables are compared with the actual calculations, it will be noted that there are slight variances in the results. These differences are due largely to (1) the rounding off of the numbers in the calculations and (2) variations in the resistance and reactance data used to prepare the tables and charts. For example, the value of the square root of three (1.732) is used frequently in three-phase calculations. Depending on the accuracy required, values of 1.7, 1.73, or 1.732 can be used in the calculations.

In actual practice, the available short-circuit current at the load side of a transformer is less than the values shown in Problem 1. However, this simplified method of finding the available short-circuit currents will result in values that are conservative and safe.

Determining the Short-circuit Current at the Terminals of a Transformer Using Tables

Figure 19-1 is a table of the short-circuit currents for a typical transformer. NEMA and

---

Determining the Short-circuit Current at the Terminals of a Transformer Using the Impedance Formula

PROBLEM 1:

Assume that the three-phase transformer installed by the utility company for the commercial building has a rating of 300 kVA at 120/208 volts with an impedance of 2% (from the transformer nameplate). The available short-circuit current at the secondary terminals of the transformer must be determined. To simplify the calculations, it is also assumed that the utility can deliver unlimited short-circuit current to the primary of the transformer. In this case, the transformer primary is known as an *infinite bus* or an *infinite primary*.

The first step is to determine the normal full-load current rating of the transformer:

$$\text{I (at the secondary)} = \frac{\text{kVA} \times 1,000}{\text{E} \times 1.73} = \frac{300 \times 1,000}{208 \times 1.73}$$

$$= 834 \text{ amperes normal full load}$$

Using the impedance value given on the nameplate of the transformer, the next step is to find a multiplier which can be used to determine the short-circuit current available at the secondary terminals of the transformer.

For a transformer impedance of 2%:

$$\frac{100}{2} = 50 \text{ (multiplier)}$$

Then,

$$834 \times 50 = 41,700 \text{ amperes of short-circuit current}$$

If the transformer impedance is 1%:

$$\frac{100}{1} = 100 \text{ (multiplier)}$$

Then,

$$834 \times 100 = 83,400 \text{ amperes of short-circuit current}$$

If the transformer impedance is 4%:

$$\frac{100}{4} = 25 \text{ (multiplier)}$$

Then,

$$834 \times 25 = 20,850 \text{ amperes of short-circuit current}$$

SYMMETRICAL SHORT-CIRCUIT CURRENTS AT VARIOUS DISTANCES FROM A LIQUID FILLED TRANSFORMER (300 kVA TRANSFORMER, 2% IMPEDANCE)

WIRE-SIZE (COPPER)

VOLTS	DIST. FT.	#14	#12	#10	#8	#6	#4	#1	0	00	000	2-000	0000	250 MCM	2-250 MCM	3-300 MCM	350 MCM	2-350 MCM	3-350 MCM	3-400 MCM	500 MCM	2-500 MCM	750 MCM	4-750 MCM
208 VOLTS	0	42090	42090	42090	42090	42090	42090	42090	42090	42090	42090	42090	42090	42090	42090	42090	42090	42090	42090	42090	42090	42090	42090	(42090)
	5	6910	10290	14730	19970	25240	29840	34690	35770	36640	37340	39610	37930	38270	40100	40870	38840	40410	40960	41030	39300	40650	39650	41460
	10	3640	5610	8460	12350	17090	22230	29030	30760	32210	33410	37340	34420	35030	38270	39710	35520	38840	39870	40010	36850	39300	37480	40840
	25	1500	2360	3670	5650	8430	12150	18930	21170	23240	25000	31710	26750	27780	33590	36560	28480	34780	36930	37230	31020	35730	32190	(39090)
	50	760	1200	1890	2950	4530	6810	11740	13670	15610	17510	25090	19320	20520	27780	32250	22660	29550	32850	33340	24520	31020	26050	36480
	100	380	600	960	1510	2350	3610	6610	7920	9320	10810	17510	12320	13380	20520	26010	15400	22660	26850	27530	17250	24520	18860	32190
	200	190	300	480	760	1190	1860	3510	4280	5140	6090	10810	7110	7860	13380	18660	9360	15400	19590	20370	10820	17250	12150	26050
	500	80	120	190	310	480	760	1460	1800	2180	2630	4990	3130	3500	6510	10030	4290	7820	10770	11400	5100	9120	5870	16570
	1000	40	60	100	150	240	380	740	910	1110	1350	2630	1620	1820	3500	5650	2250	4290	6140	6560	2710	5100	3160	10310
	5000	10	10	20	30	50	80	150	180	230	280	550	330	380	740	1260	470	930	1380	1490	570	1130	670	2560
240 VOLTS	0	37820	37820	37820	37820	37820	37820	37820	37820	37820	37820	37820	37820	37820	37820	37820	37820	37820	37820	37820	37820	37820	37820	37820
	5	7750	11330	15810	20720	25260	28940	32560	33340	33960	34460	36080	34870	35120	36420	36960	35520	36640	37020	37070	35840	36800	36090	37370
	10	4140	6320	9400	13430	18040	22670	28230	29560	30660	31550	34460	32290	32730	35120	36140	33470	35520	36260	36350	34060	35840	34510	36930
	25	1720	2700	4180	6360	9360	13190	19640	21620	23380	24920	30240	26260	27090	31650	33860	28480	32530	34130	34340	29610	33230	30510	35680
	50	870	1380	2160	3360	5130	7620	12730	14630	16480	18230	24920	19850	20900	27090	30610	22740	28480	31060	31430	24300	29610	25570	33780
	100	440	700	1100	1730	2680	4100	7380	8770	10220	11720	18230	13220	14240	20900	25600	16150	22740	26280	26830	17860	24300	19310	30510
	200	220	350	550	880	1370	2130	3990	4830	5770	6790	11720	7880	8650	14240	19200	10190	16150	20030	20710	11650	17860	12960	25570
	500	90	140	220	350	560	870	1670	2050	2490	2990	5610	3540	3960	7230	10890	4820	8590	11620	12250	5700	9920	6520	16890
	1000	40	70	110	180	280	440	850	1050	1280	1550	2990	1850	2080	3960	6300	2560	4820	6820	7270	3070	5700	3570	11130
	5000	10	10	20	40	60	90	170	210	260	320	630	380	430	860	1440	540	1070	1580	1710	660	1290	770	2910
480 VOLTS	0	18910	18910	18910	18910	18910	18910	18910	18910	18910	18910	18910	18910	18910	18910	18910	18910	18910	18910	18910	18910	18910	18910	18910
	5	10450	12820	14750	16150	17080	17690	18200	18310	18400	18470	18690	18520	18550	18730	18800	18610	18760	18810	18810	18650	18780	18690	18850
	10	6750	9170	11630	13780	15400	16530	17540	17740	17910	18040	18470	18150	18210	18550	18690	18320	18610	18710	18720	18400	18650	18470	18800
	25	3180	4740	6770	9150	11520	13570	15690	16160	16540	16840	17830	17100	17250	18040	18380	17490	18180	18410	18440	17690	18280	17840	18630
	50	1680	2590	3900	5680	7840	10170	13190	13960	14600	15120	16840	15560	15820	17250	17870	16260	17490	17940	18000	16610	17690	16890	18360
	100	860	1350	2090	3180	4680	6600	9820	10810	11690	12460	15120	13130	13540	15820	16930	14240	16260	17060	17170	14810	16610	15260	17840
	200	440	690	1080	1680	2560	3810	6370	7320	8240	9110	12460	9930	10450	13540	15300	11370	14240	15530	15710	12150	14810	12780	16890
	500	180	280	440	700	1080	1670	3040	3640	4290	4960	8010	5560	6140	9360	11820	7050	10320	12190	12500	7880	11150	8600	14550
	1000	90	140	220	350	550	860	1620	1970	2370	2800	4960	3270	3610	6140	8520	4300	7050	8940	9290	4960	7880	5560	11830
	5000	20	30	40	70	110	180	340	420	510	620	1210	750	840	1610	2600	1040	1980	2820	3020	1250	2350	1450	4730
600 VOLTS	0	15130	15130	15130	15130	15130	15130	15130	15130	15130	15130	15130	15130	15130	15130	15130	15130	15130	15130	15130	15130	15130	15130	15130
	5	10210	11790	12920	13690	14180	14500	14770	14820	14870	14900	15010	14930	14940	15040	15070	14970	15050	15080	15080	15000	15060	15010	15100
	10	7270	9270	11010	12350	13280	13890	14410	14520	14610	14680	14900	14730	14770	14940	15020	14820	14970	15020	15030	14870	15000	14900	15070
	25	3740	5370	7280	9230	10920	12200	13410	13670	13870	14040	14570	14170	14250	14680	14850	14380	14750	14870	14890	14490	14800	14570	14980
	50	2040	3080	4500	6270	8170	9950	11940	12400	12770	13060	14040	13310	13460	14250	14590	13700	14380	14620	14650	13900	14490	14050	14840
	100	1060	1650	2510	3730	5290	7080	9650	10350	10930	11420	13060	11840	12090	13460	14080	12510	13700	14150	14210	12850	13900	13120	14570
	200	540	850	1330	2040	3040	4390	6840	7640	8390	9050	11420	9640	10010	12090	13160	10640	12510	13290	13390	11160	12850	11580	14050
	500	220	350	550	860	1330	2010	3550	4180	4830	5480	8180	6110	6530	9210	10960	7310	9900	11210	11400	7990	10470	8560	12700
	1000	110	170	280	440	680	1050	1950	2360	2800	3270	5480	3760	4110	6530	8540	4780	7310	8860	9120	5410	7990	5970	10940
	5000	20	40	60	90	140	220	420	520	640	770	1470	910	1030	1930	3030	1260	2340	3270	3480	1510	2750	1740	5180

Fig. 19-1

transformer manufacturers publish short-circuit tables for many sizes of transformers having various impedance values. The table in figure 19-1 provides data for a 300-kVA, three-phase transformer with an impedance of 2%. According to the table, the symmetrical short-circuit current is 42,090 amperes at the secondary terminals of a 120/208-volt transformer (refer to the zero-foot row of the table). This value corresponds very closely to the value of 41,700 amperes obtained from the use of the Impedance Formula. The table shown in figure 19-2 also indicates that the short-circuit current is 41,700 amperes.

Determining the Short-circuit Current at Various Distances from a Transformer Using the Table in Figure 19-1.

The amount of available short-circuit current decreases as the distance from the transformer increases, as indicated in figure 19-1. See Problem 2.

Determining Short-circuit Currents at Various Distances from Transformers, Switchboards, Panelboards, and Load Centers Using the Point-to-point Method

A simple method of determining the available short-circuit currents at various distances from a given location is the *point-to-point method*. Reasonable accuracy is obtained when this method is used with three-phase and single-phase systems.

The following procedure demonstrates the use of the point-to-point method.

Step 1. Determine the full-load rating of the transformer in amperes from the transformer nameplate, tables, or the following formulas:

a. For three-phase transformers:

$$I_{FLA} = \frac{kVA \times 1,000}{E_{L-L} \times 1.73}$$

where E_{L-L} = Line-to-line voltage

b. For single-phase transformers:

$$I_{FLA} = \frac{kVA \times 1,000}{E_{L-L}}$$

Step 2. Find the transformer multiplier:

$$Multiplier = \frac{100}{transformer \% \text{ impedance } (Z)}$$

Step 3. Determine the transformer let-through short-circuit current from tables or use the following formula:

$$I_{SCA} = transformer\ FLA \times multiplier$$

Step 4. Determine the f factor:

a. For three-phase faults: $f = \dfrac{1.73 \times L \times I}{C \times E_{L-L}}$

b. For single-phase, line-to-line (L-L) faults on single-phase, center-tapped transformers:

$$f = \frac{2 \times L \times I}{C \times E_{L-L}}$$

c. For single-phase, line-to-neutral (L-N) faults on single-phase, center-tapped transformers:

$$f = \frac{2 \times L \times I^*}{C \times E_{L-N}}$$

where

L = the length of the circuit to the fault, in feet (meters).

I = the available short-circuit current at the beginning of the circuit, in amperes.

C = the constant derived from the tables in figure 19-3 for the specific type of conductors and circuit arrangement. For parallel runs, multiply the C value by the number of conductors per phase.

E = the voltage, line-to-line or line-to-neutral.

*I = 1.5 × L-L short-circuit current in amperes at the transformer terminals. The L-N fault current is higher than the L-L fault current at the secondary terminals of a single-phase, center-tapped transformer. At some distance from the terminals, depending upon the wire size, the L-N fault

---

PROBLEM 2:

For a 300-kVA transformer with a secondary voltage of 208 volts, find the available short-circuit current at a main switch which is located 25 feet from the transformer. The main switch is supplied by four No. 750 MCM copper conductors per phase in steel conduit.

Refer to the table in figure 19-1 and read the value of 39,090 amperes in the column on the right-hand side of the table for a distance of 25 feet.

KVA	Full Load Amps	Voltage	Phase	Percent Impedance	Short-Circuit‡ Amps
25	104	120/240	1	1.6	10,300 *
37 1/2	156	120/240	1	1.6	15,280 *
50	209	120/240	1	1.7	19,050 *
75	313	120/240	1	1.6	29,540 *
100	417	120/240	1	1.6	38,540 *
167	695	120/240	1	1.8	54,900 *
150	417	120/208	3	2	20,850
225	625	120/208	3	2	31,250
300	834	120/208	3	2	41,700
500	1388	120/208	3	2	69,400
750	2080	120/208	3	5**	41,600
1000	2776	120/208	3	5**	55,520
1500	4164	120/208	3	5**	83,280
2000	5552	120/208	3	5**	111,040
2500	6950	120/208	3	5**	139,000
150	181	277/480	3	2	9,050
225	271	277/480	3	2	13,550
300	361	277/480	3	2	18,050
500	601	277/480	3	2	30,050
750	902	277/480	3	5**	18,040
1000	1203	277/480	3	5**	24,060
1500	1804	277/480	3	5**	36,080
2000	2406	277/480	3	5**	48,120
2500	3007	277/480	3	5**	60,140

† 3∅ short-circuit currents based on infinite primary. 1-∅ short-circuit currents based on 100,000 kVA primary.

‡ These values of short-circuit current may be higher if a lower percent impedance transformer is used. Consult your local utility.

* Values based on maximum short-circuit current condition at single-phase transformer secondary terminals. Maximum short-circuit current will occur across the 120-volt transformer terminals.

** These transformers are available in lower impedances, resulting in greater short-circuit currents. It is important that all transformer impedances be checked before determining interrupting requirements.

NOTE: An interrupting capacity rating for an overcurrent protective device is a maximum rating and should not be exceeded.

Fig. 19-2 Tables of short-circuit currents available from various size transformers. †

| AWG or MCM | Copper Three Single Conductors | | | | Copper Three-Conductor Cable | | Aluminum Three Single Conductors or Three-Conductor Cables | | AWG or MCM |
| | In Magnetic Duct | | In Nonmagnetic Duct | | In Magnetic Duct | In Nonmagnetic Duct | In Magnetic Duct | In Nonmagnetic Duct | |
	600 V and 5 kv Nonshielded	5 kv Shielded and 15 kv	600 V and 5 kv Nonshielded	5 kv Shielded and 15 kv	600 V and 5 kv Nonshielded	600 V and 5 kv Nonshielded	600 V and 5 kv Nonshielded	600 V and 5 kv Nonshielded	
12	617	–	–	–	–	–	–	–	12
10	982	–	–	–	–	–	–	–	10
8	1230	1230	1230	1230	1230	1230	–	–	8
6	1940	1940	1950	1940	1950	1950	1180	1180	6
4	3060	3040	3080	3070	3080	3090	1870	1870	4
3	3860	3830	3880	3870	3880	3900	2360	2360	3
2	4760	4670	4830	4780	4830	4850	2960	2970	2
1	5880	5750	6020	5920	6020	6100	3720	3750	1
1/0	7190	6990	7460	7250	7410	7580	4670	4690	1/0
2/0	8700	8260	9090	8770	9090	9350	5800	5880	2/0
3/0	10400	9900	11500	10700	11100	11900	7190	7300	3/0
4/0	12300	10800	13400	12600	13400	14000	8850	9170	4/0
250	13500	12500	14900	14000	14900	15800	10300	10600	250
300	14800	13600	16700	15500	16700	17900	11900	12400	300
350	16200	14700	18700	17000	18600	20300	13500	14200	350
400	16500	15200	19200	17900	19500	21100	14800	15800	400
450	17300	15900	20400	18800	20700	22700	–	–	450
500	18100	16500	21500	19700	21900	24000	17200	18700	500
600	18900	17200	22700	20900	23300	25700	18900	21000	600
700	–	–	–	–	–	–	20500	23100	700
750	20200	18300	24700	22500	25600	28200	21500	24300	750
1000	–	–	–	–	–	–	23600	27600	1000

| Ampacity | Plug-in Busway | | Feeder Busway | | High Impedance Busway |
	Copper	Aluminum	Copper	Aluminum	Copper
225	28700	23000	18700	12000	–
400	38900	34700	23900	21300	–
600	41000	38300	36500	31300	–
800	46100	57500	49300	44100	–
1000	69400	89300	62900	56200	15600
1200	94300	97100	76900	69900	16100
1350	119000	104200	90100	84000	17500
1600	129900	120500	101000	90900	19200
2000	142900	135100	134200	125000	20400
2500	143800	156300	180500	166700	21700
3000	144900	175400	204100	188700	23800
4000	–	–	277800	256400	–

Fig. 19-3 Tables of C values.

current is lower than the L-L fault current. The 1.5 multiplier is an approximation and can vary from 1.33 to 1.67. These figures are based on the change in the turns ratio between the primary and secondary, infinite source impedance, a distance of zero feet from the terminals of the transformer, and 1.2 X % reactance (X) and 1.5 X % resistance (R) for the L-N vs. L-L resistance and reactance values. Begin the L-N calculations at the transformer secondary terminals and then proceed with the point-to-point method.

Step 5. After finding the f factor, refer to figure 19-4 and locate in Chart M the appropriate value of the multiplier (M) for the specific f value.

Step 6. Multiply the available short-circuit current at the beginning of the circuit by the multiplier (M) to determine the available symmetrical short-circuit current at the fault.

I_{SCA} at fault = I_{SCA} at beginning of circuit X M

CHART M (multiplier)

f	M	f	M
0.01	0.99	1.20	0.45
0.02	0.98	1.50	0.40
0.03	0.97	2.00	0.33
0.04	0.96	3.00	0.25
0.05	0.95	4.00	0.20
0.06	0.94	5.00	0.17
0.07	0.93	6.00	0.14
0.08	0.93	7.00	0.13
0.09	0.92	8.00	0.11
0.10	0.91	9.00	0.10
0.15	0.87	10.00	0.09
0.20	0.83	15.00	0.06
0.30	0.77	20.00	0.05
0.40	0.71	30.00	0.03
0.50	0.67	40.00	0.02
0.60	0.63	50.00	0.02
0.70	0.59	60.00	0.02
0.80	0.55	70.00	0.01
0.90	0.53	80.00	0.01
1.00	0.50	90.00	0.01
		100.00	0.01

$$M = \frac{1}{1 + f}$$

Fig. 19-4

Motor Contribution. All motors running at the instant a short circuit occurs contribute to the short-circuit current. The amount of current from the motors is equal approximately to the starting (locked rotor) current for each motor. This current value depends upon the type of motor, its characteristics, and its code letter. Refer to *NEC Table 430-7(b)*. It is a common practice to multiply the full-load ampere rating of the motor by 4 or 5 to obtain a close approximation of the locked rotor current and provide a margin of safety. The current contributed by running motors at the instant a short circuit occurs is added to the value of the short-circuit current at the main switchboard prior to the start of the point-to-point method. To simplify the following problems, motor contributions are not added to the short-circuit currents.

Note: A three-phase fault current value determined in Step 6 is the approximate current that will flow if the three hot conductors of a three-phase system are shorted together.

To obtain approximate short-circuit current values when two hot conductors of a three-phase

system are shorted together, use 86% of the three-phase fault current value. In other words, if the three-phase current value is 20,000 amperes when the three lines are shorted together (L-L-L value), then the short-circuit current for two shorted lines (L-L value) is approximately 20,000 X 0.86 = 17,200 amperes. A value of 58% of the three-phase fault current is used to approximate the short-circuit current value when one hot conductor and the neutral (ground) of a three-phase system are shorted together (L-N). That is, if the three-phase (L-L-L) value is 20,000 amperes, then the L-N short-circuit current is approximately 20,000 X 0.58 = 11,600 amperes.

As shown in Problem 3, the fuses or circuit breakers located in the main switchboard of the commercial building must have an interrupting capacity of at least 35,862 amperes rms symmetrical. It is good practice to install protective devices having an interrupting rating at least 25% greater than the actual calculated available fault current. This practice generally provides a margin of safety to permit the rounding off of numbers, as well as compensating for a reasonable amount of short-circuit contribution from any electrical motors that may be running at the instant the fault occurs.

The fuses specified for the commercial building have an interrupting rating of 200,000 amperes (see the Specifications). In addition, the switchboard bracing is specified to be 50,000 amperes.

If current-limiting fuses are installed in the main switchboard feeders protecting the various panelboards, breakers having values of 5,000 or 10,000 AIC may be installed in the panelboards. This installation meets the requirements of *NEC Section 110-9.* (If necessary, the student should review the sections covering peak let-through charts in Unit 18.)

If noncurrent-limiting overcurrent devices (standard molded case circuit breakers) are to be installed in the main switchboard, breakers having adequate interrupting ratings must be installed in the panelboards. A short-circuit study must be

---

PROBLEM 3:

It is desired to find the available short-circuit current at the main switchboard of the commercial building. Once this value is known, the electrician can provide overcurrent devices with adequate interrupting capacities and the proper bus bar bracing within the switchboard (see *NEC Sections 110-9, 110-10 and 230-98*). Figure 19-5 shows the actual electrical system for the commercial building. As each of the following steps in the point-to-point method is examined, refer to the tables given in figures 19-1 through 19-4 to locate the necessary values of C, f, and M.

Step 1. $I_{FLA} = \dfrac{kVA \times 1{,}000}{E_{L\text{-}L} \times 1.73} = \dfrac{300 \times 1{,}000}{208 \times 1.73} = 834$ amperes

Step 2. Multiplier $= \dfrac{100}{\text{Trans. } \% Z} = \dfrac{100}{2} = 50$

Step 3. $I_{SCA} = 834 \times 50 = 41{,}700$ amperes

Step 4. $f = \dfrac{1.73 \times L \times 1}{3 \times C \times E_{L\text{-}L}} = \dfrac{1.73 \times 25 \times 41{,}700}{3 \times 18{,}100 \times 208} = 0.16$

Step 5. $M = \dfrac{1}{1+f} = \dfrac{1}{1+0.16} = 0.86$ (See figure 19-4)

Step 6. The short-circuit available at the line side lugs on the main switchboard is:

$41{,}700 \times 0.86 = 35{,}862$ amperes rms symmetrical

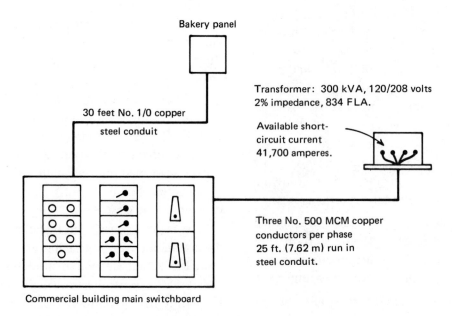

Bakery panel

30 feet No. 1/0 copper

steel conduit

Transformer: 300 kVA, 120/208 volts 2% impedance, 834 FLA.

Available short-circuit current 41,700 amperes.

Three No. 500 MCM copper conductors per phase 25 ft. (7.62 m) run in steel conduit.

Commercial building main switchboard

Fig. 19-5

PROBLEM 4. Single-phase Transformer:

This problem is illustrated in figure 19-6. The point-to-point method is used to determine the currents for both line-to-line and line-to-neutral faults for a 167-kVA, 2% impedance transformer on a 120/240-volt, single-phase system.

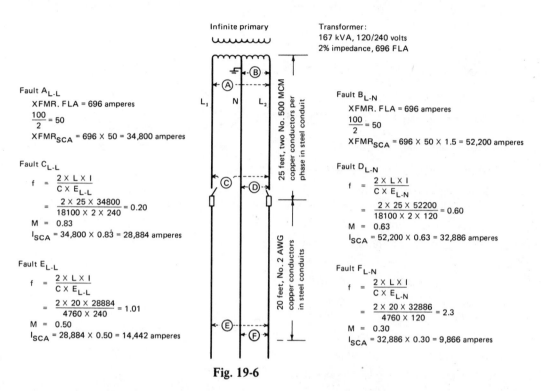

Fig. 19-6

made for each panelboard location to determine the value of the available short-circuit current.

The cost of circuit breakers increases as the interrupting rating of the breaker increases. The most economical protection system results when current-limiting fuses are installed in the main switchboard to protect the breakers in the panelboards. In this case, the breakers in the panelboards will have the standard 5,000 or 10,000 AIC ratings.

Summary

To meet the requirements of *NEC Sections 110-9, 110-10*, and *230-98*, three methods can be used to determine the necessary overcurrent protective devices having adequate ratings for given amounts of available short-circuit current.

1. For an installation using fuses only, the fuses must have an interrupting rating *not less* than the available fault current. It is recommended that the fuses used have interrupting ratings *at least 25% greater* than the available fault current.

2. For an installation using circuit breakers only, the breakers must have an interrupting rating *not less* than the available fault current at the point of application. It is recommended that the breakers used have interrupting ratings *at least 25% greater* than the available fault current.

3. Install fusible mains, feeders, and motor controllers. Provide the proper current-limiting fuses to protect panels containing circuit breakers which have inadequate interrupting ratings. Use peak let-through charts or fuse-breaker application tables from manufacturers to select current-limiting fuses.

COORDINATION OF OVERCURRENT PROTECTIVE DEVICES

While this text cannot cover the topic of electrical system coordination (selectivity) in detail,

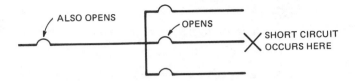

THE FAULT ON THE BRANCH CIRCUIT TRIPS BOTH THE
BRANCH-CIRCUIT BREAKER AND THE FEEDER CIRCUIT
BREAKER. AS A RESULT, POWER TO THE PANEL IS CUT
OFF AND CIRCUITS THAT SHOULD NOT BE AFFECTED
ARE NOW OFF.

Fig. 19-7 Nonselective system.

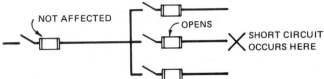

ONLY THE BRANCH-CIRCUIT FUSE OPENS. ALL
OTHER CIRCUITS AND THE MAIN FEEDER RE-
MAIN ON.

Fig. 19-8 Selective system.

the following material will provide the student
with a working knowledge of this important topic.

What is Coordination? (*Section 240-12*)

A situation known as *nonselective coordination* occurs when a fault on a branch circuit opens
not only the branch-circuit overcurrent device, but
also opens the feeder overcurrent device, figure 19-
7. Nonselective systems are installed unknowingly
and cause needless power outages in portions of an
electrical system that should not be affected by a
fault.

A *selectively coordinated system*, figure 19-8,
is one in which *only* the overcurrent device imme-
diately upstream from the fault opens. Obviously,
the installation of a selective system is much more
desirable than a nonselective system.

The importance of selectivity in an electrical
system is covered extensively throughout *NEC Article 517*. This article pertains to health care facilities,
where maintaining electrical power is extremely
important. The unexpected loss of power in certain
areas of hospitals, nursing care centers, and similar
health care facilities can be catastrophic.

Some local electrical codes require that *all*
circuits, feeders, and mains in health care facilities
be designed so as to be *selective* in nature. The
local electrical code should be consulted.

The importance of selectivity (system coordination) is also emphasized in Code *Section 230-95(a),
Exception No. 1*; and in *Section 240-12*. Both sections refer to industrial installations where additional
hazards would be introduced should a nonorderly
shutdown occur. The Code defines coordination,
in part, as the proper localizing of a fault condition
to restrict outages to the equipment affected.

By knowing how to determine the available
short-circuit current and ground-fault current the

electrician can make effective use of the time-
current curves and peak let-through charts (Unit
18) to find the length of time required for a fuse to
open or a circuit breaker to trip.

What Causes Nonselectivity?

In figure 19-9, a short circuit in the range
of 3,000 amperes occurs on the load side of a
20-ampere breaker. The magnetic trip of the
breaker is adjusted permanently by the manufacturer to unlatch at a current value equal to 10
times its rating or 200 amperes. The feeder breaker
is rated at 100 amperes; the magnetic trip of this
breaker is set by the manufacturer to unlatch at a
current equal to 10 times its rating or 1,000
amperes. This type of breaker generally cannot be
adjusted in the field. Therefore, a current of 200
amperes or more will cause the 20-ampere breaker
to trip instantly. In addition, any current of 1,000
or more will cause the 100-ampere breaker to trip
instantly.

For the breakers shown in figure 19-9, a momentary fault of 3,000 amperes will trip (unlatch)
both breakers. Since the flow of current in a series
circuit is the same in all parts of the circuit, the
3,000-ampere fault will trigger both magnetic trip
mechanisms. The time-current curve shown in
figure 18-27 indicates that for a 3,000-ampere
fault, the unlatching time for both breakers is
0.0042 seconds and the interrupting time for both
breakers is 0.016 seconds.

This example of a nonselective system should
make apparent to the student the need for a
thorough study and complete understanding of
time-current curves, fuse selectivity ratios, and
unlatching time data for circuit breakers. Otherwise, a blackout may occur, such as the loss of exit
and emergency lighting. The student must be able

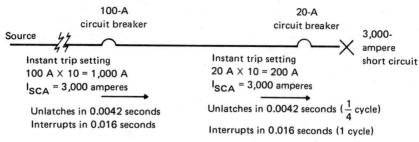

Fig. 19-9 Nonselective system.

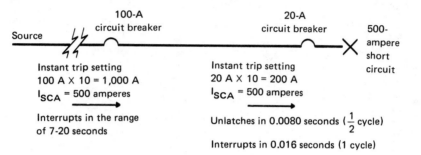

Fig. 19-10 Selective system.

to determine available short-circuit currents (1) to insure the proper selection of protective devices with adequate interrupting ratings and (2) to provide the proper coordination as well.

Figure 19-10 shows an example of a selective circuit. In this circuit, a fault current of 500 amperes trips the 20-ampere breaker instantly (the unlatching time of the breaker is approximately 0.0080 seconds and its interrupting time is 0.016 seconds). The graph in figure 18-27 indicates that the 100-ampere breaker interrupts the 500-ampere current in a range from 7 to 20 seconds. This relatively lengthy trip time range is due to the fact that the 500-ampere fault acts upon the current thermal trip element only, and does not affect the magnetic trip element which operates on a current of 1,000 amperes or more.

Selective System Using Fuses

The proper choice of the various classes and types of fuses is necessary if selectivity is to be achieved, figure 19-11. Indiscriminate mixing of fuses of different classes, time-current characteristics, and even manufacturers may cause a system to become nonselective.

To insure selective operation under low-overload conditions, it is necessary only to check and compare the time-current characteristic curves of fuses. Selectivity occurs when the curves do not cross one another.

Fuse manufacturers publish *Selectivity Guides*, similar to the one shown in figure 19-12, to be used for short-circuit conditions. When using these guides, selectivity is achieved by maintaining a specific amperage ratio between the various classes and types of fuses. A selectivity chart is based on any fault current up to the maximum interrupting ratings of the fuses listed in the chart.

SINGLE PHASING

The *National Electrical Code* requires that all three-phase motors be provided with running overcurrent protection in each phase. A line-to-ground fault generally will blow one fuse. There will be a resulting increase of 173–200% in the line current in the remaining two connected phases. This increased current will be sensed by the motor fuses and overload relays when such fuses and relays are sized at 125% or less of the full-load current rating of the motor. Thus, the fuses and/or the overload relays will open before the motor windings are damaged.

A line-to-line fault in a three-phase motor will blow two fuses. In general, the operating coil of

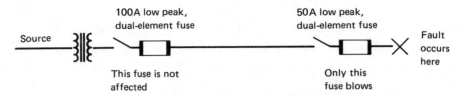

Fig. 19-11 Selective system using fuses.

RATIOS FOR SELECTIVITY

LINE SIDE FUSE	LOAD SIDE FUSE →								
	KRP-C HI-CAP time-delay Fuse 601-6000A Class L	KTU LIMITRON fast-acting Fuse 601-6000A Class L	KTN-R, KTS-R LIMITRON fast-acting Fuse 0-600A Class RK1	JJS, JJN TRON fast-acting Fuse (Class T) 0-600A	JKS LIMITRON quick-acting Fuse (Class J) 0-600A	FRN-R, FRS-R FUSETRON dual-element Fuse 0-600A Class RK5	LPN-R, LPS-R LOW-PEAK dual-element Fuse 0-600A Class RK1	JHC HI-CAP time-delay Fuse (Class J Dim) 15-600A	SC Type Fuse (Class G) 0-60A
KRP-C HI-CAP time-delay Fuse 601-6000A Class L	2:1	2:1	2:1	2:1	2:1	4:1	2:1	3:1	**
KTU LIMITRON fast-acting Fuse 601-6000A Class L	2:1	2:1	2:1	2:1	2:1	6:1	2:1	5:1	**
KTN-R, KTS-R LIMITRON fast-acting Fuse 0-600A Class RK1	N/A	N/A	3:1	3:1	3:1	8:1	3:1	4:1	4:1
JJN, JJS TRON fast-acting Fuse Class T 0-600A	N/A	N/A	3:1	3:1	3:1	8:1	3:1	4:1	4:1
JKS LIMITRON quick-acting Fuse Class J 0-600A	N/A	N/A	3:1	3:1	3:1	8:1	3:1	4:1	4:1
FRN-R, FRS-R, FUSETRON dual-element Fuse 0-600A Class RK5	N/A	N/A	1.5:1	1.5:1	1.5:1	2:1	1.5:1	1.5:1	1.5:1
LPN-R, LPS-R LOW-PEAK dual-element Fuse 0-600A Class RK1	N/A	N/A	1.5·1	1.5:1	1.5:1	4:1	2:1	2:1	2:1
JHC HI-CAP time-delay Fuse (Class J Dim.) 15-600A	N/A	N/A	1.5:1	1.5:1	1.5:1	4:1	2:1	2:1	2:1
SC Type Fuse Class G 0-60A	N/A	N/A	2:1	2:1	2:1	4:1	3:1	3:1	2:1

*Applies only to the indicated BUSS fuses.

**SC Fuses available in sizes up to 60 amperes. Selectivity is not an issue here.

N/A — Not applicable. A line side fuse rated less than load size fuse not normally installed.

NOTE: At some values of fault current these ratios may be lowered to permit closer fuse sizing. Check with manufacturer for lower ratio possibilities or plot fuse curves.

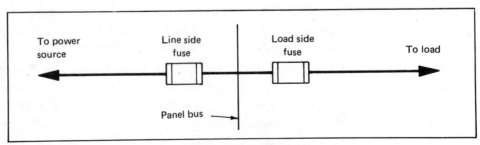

Fig. 19-12 Selectivity guide.

PROBLEM 5

It is desired to install 100-ampere, low-peak, dual-element fuses in a main switch, and a 50-ampere, low-peak, dual-element fuse in the feeder switch. Is this combination of fuses selective?

Refer to the Selectivity Guide in figure 19-12. The chart indicates that a 2:1 minimum ratio must be maintained for this type of fuse. Since 100:50 is a 2:1 ratio, the installation is selective. In addition, any fuse of the same type having a rating of less than 50 amperes will also be selective with the 100-ampere main fuse. That is, if a fault occurs on the load side of the 50-ampere low-peak fuse, only the 50-ampere fuse will open.

the motor controller will drop out, thus providing protection to the motor winding.

To reduce *single-phasing* problems, each three-phase motor must be provided with individual protection through the proper sizing of the overload relays and fuses. Phase failure relays are also available. However, can other equipment be affected by a single-phasing condition?

In general, loads that are connected line-to-neutral or line-to-line, such as lighting, receptacles, and electric heating units are not affected by a single-phasing condition. In other words, if one main fuse blows, then two-thirds of the

lighting, receptacles, and electric heat will remain on. If two main fuses blow, then one-third of the lighting remains on, and a portion of the electric heat connected line-to-neutral will stay on.

It is essential to maintain some degree of lighting in occupancies such as stores, schools, offices, and health care facilities (such as nursing homes). A total blackout in these public structures has the potential for causing extensive personal injury due to panic. A loss of one or two phases of the system supplying a building should not cause a complete power outage in the building.

—REVIEW—

Note: Refer to the *National Electrical Code* or the plans as necessary.

1. Using the table in figure 19-1, determine the available short-circuit currents on a 208-volt system for the following:
   a. 50 feet (15.24 m) of No. 1 conductor _____ A
   b. 25 feet (7.62 m) of No. 3/0 conductor _____ A
   c. 50 feet (15.24 m) of No. 500 MCM conductor _____ A
   d. 50 feet (15.24 m) of two No. 3/0 conductors per phase _____ A
   e. 100 feet (30.48 m) of No. 4/0 conductor _____ A

2. a. Define *selectivity* _____
   _____
   _____
   _____
   _____

   b. Define *nonselectivity* _____
   _____
   _____
   _____
   _____

3. Indicate whether the following systems are selective or nonselective.
   a. A KRP-C 2,000-ampere fuse installed ahead of an LPS-R 600-ampere fuse.

   _____

   b. An FRS-R 600-ampere Fusetron fuse installed ahead of an LPS-R 400-ampere low-peak fuse. _____
   c. A KTS-R 400-ampere Limitron fuse installed ahead of an FRS-R 200-ampere Fuse-tron fuse. _____

4. Refer to figure 19-13 and calculate the available short-circuit current at Panel "A." This panel is supplied by a 25-foot (7.62 m) run of No. 3/0 copper conductors in steel conduit. Use the point-to-point method and show all calculations.

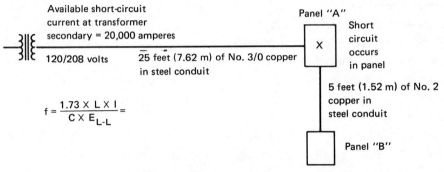

Fig. 19-13

(STUDENT CALCULATIONS)

5. Calculate the available fault current if a short circuit occurs in Panel B, figure 19-13. Use the point-to-point method and show all calculations.

(STUDENT CALCULATIONS)

6. Calculate the available fault current if a short circuit occurs in the bakery panel of the commercial building covered in this text. See figure 19-5. Show all calculations.

(STUDENT CALCULATIONS)

UNIT 20

Panelboard Selection and Installation

OBJECTIVES

After completing the study of this unit, the student will be able to

- select a panelboard, given the number of required branch circuits and their load.
- connect the branch circuits to a panelboard.
- select the feeder size, given the loading of the panelboard.

Changing needs for electrical energy in a commercial building will require the electrician to install new branch circuits. In addition, as the existing panelboards are filled, new panelboards must be installed. As a result, the electrician must be able to select and connect panelboards and the feeders to supply them.

PANELBOARDS

Separate feeders must be run from the main service equipment to each of the areas of the commercial building. Each feeder will terminate in a panelboard which is located in the area to be served, figure 20-1.

The *National Electrical Code* defines a panelboard as a single panel or group of panel units designed for assembly in the form of a single panel; such a panel will include buses, automatic overcurrent devices, and may or may not include switches for the control of light, heat, or power circuits. A panelboard is designed to be installed in a cabinet or cutout box placed in or against a wall or partition. This cabinet (and panelboard) is to be accessible only from the front. Panelboards shall be dead front (*Section 384-18*).

While a panelboard is accessible only from the front, a switchboard may be accessible from the rear as well. Note that the *NEC* in *Section 384-14*

further defines a lighting and appliances branch-circuit panelboard as a panelboard having more than 10 percent of its overcurrent devices rated at 30 amperes or less with neutral connections provided for these devices. The maximum number of overcurrent devices in a panelboard is 42.

Lighting and appliance branch-circuit panelboards generally are used on four-wire, 120/208-volt electrical systems. This type of installation provides individual branch circuits for lighting and receptacle outlets and permits the connection of appliances such as the equipment in the bakery.

Number of Overcurrent Devices

Not more than 42 overcurrent devices are permitted in a lighting and appliance branch-circuit panelboard, according to *NEC Section 384-15*. The actual number of overcurrent devices is determined from the needs of the area to be served. Referring to the panel schedule for the bakery (following the Specifications provided after this unit), note that 24 poles are shown: 17 poles have connected loads, 3 are spares and 4 are spaces. This value of 24 poles was determined by the actual needs and then a reasonable number of circuits was added for growth possibilities. Since panelboards usually are available in multiples of the phase connections, the number of poles for this installation is selected to

Fig. 20-1 Panelboards.

be equal to 3 X 8 or 24 poles. This installation will allow for future expansion.

As discussed in Unit 18, circuit breakers shall not be used as switches unless marked *SWD*.

Panelboard Construction

In general, panelboards are constructed so that the main feed bus bars run the height of the panelboard. The buses to the branch-circuit protective devices are connected to the alternate main buses as shown in figures 20-2 and 20-3. In an arrangement of this type, the connections directly across from each other are on the same phase and the adjacent connections on each side are on different phases. As a result, multiple protective devices can be installed to serve the 208-volt equipment. The numbering sequence shown in figures 20-2 and 20-3 is the common numbering system used for most panels. Figures 20-4 and 20-5 show the phase arrangement requirements of *NEC Section 384-3*.

Color Code for Branch Circuits

The hot ungrounded conductors may be any color *except* GREEN — which is reserved for grounding purposes only, or WHITE or GRAY — which are reserved for the grounded circuit conductor.

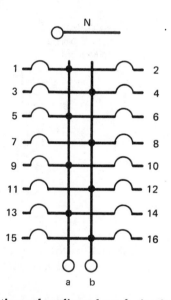

Fig. 20-2 Lighting and appliance branch-circuit panelboard; single-phase, three-wire connections.

Panelboard Sizing

Panelboards are available in various sizes including 100, 225, 400 and 600 amperes. These sizes are determined by the current-carrying capacity of the main bus. After the feeder capacity is determined, the panelboard size is selected to be not smaller than the feeder size (refer to *NEC Section 384-13*).

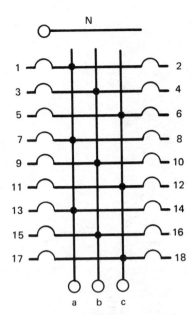

Fig. 20-3 Lighting and appliance branch-circuit panelboard; three-phase, four-wire connections.

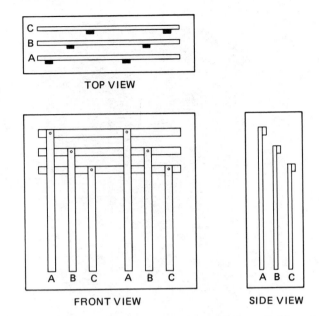

TOP VIEW

FRONT VIEW   SIDE VIEW

Fig. 20-4 Phase arrangement requirements for switchboards and panelboards, according to *NEC Section 384-3*.

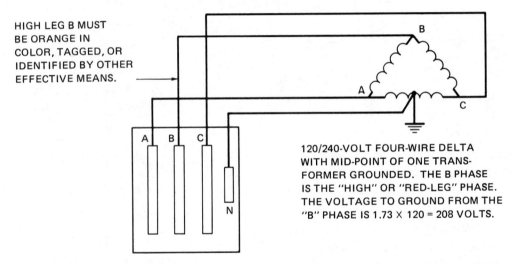

HIGH LEG B MUST BE ORANGE IN COLOR, TAGGED, OR IDENTIFIED BY OTHER EFFECTIVE MEANS.

120/240-VOLT FOUR-WIRE DELTA WITH MID-POINT OF ONE TRANS-FORMER GROUNDED. THE B PHASE IS THE "HIGH" OR "RED-LEG" PHASE. THE VOLTAGE TO GROUND FROM THE "B" PHASE IS 1.73 X 120 = 208 VOLTS.

Fig. 20-5 Panelboards and switchboards supplied by four-wire delta-connected systems shall have the "B" phase connected to the phase having the higher voltage to ground. See *NEC Section 384-3(f)*.

Panelboard Overcurrent Protection

In many installations, a single feeder may be sized to serve several panelboards, figure 20-6. For this situation, it is necessary to install an overcurrent device with a trip rating not greater than the rating of the bus bar in each panelboard, figure 20-7. In the case of the bakery installation, a main device is not required as the panelboard is protected by the feeder protective device. Other examples of methods of providing panelboard protection are shown in figures 20-8 and 20-9.

THE FEEDER

The size of a feeder is determined by the branch-circuit loads to be supplied. In addition, a reasonable allowance is made for future electrical needs. It is accepted practice to allow a growth factor of 25%. A practical method of determining the feeder sizing is to assume that the feeder is continuously loaded and a major portion of the load consists of electric-discharge lighting.

As an example, consider the bakery loading:

33,900 volt-amps ÷ (208 X 1.73) volts = 94 amperes

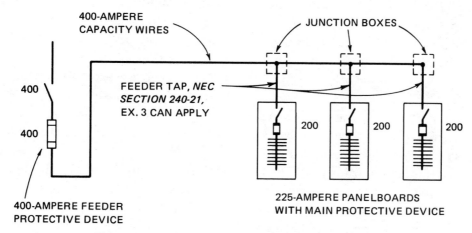

Fig. 20-6 Panelboards with main. *NEC Section 384-16.*

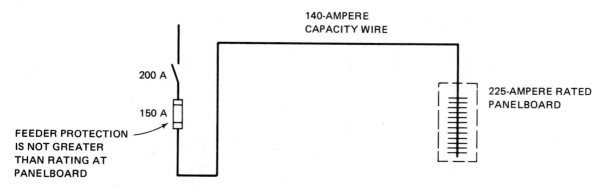

Fig. 20-7 Panelboard without main. *NEC Section 384-16.*

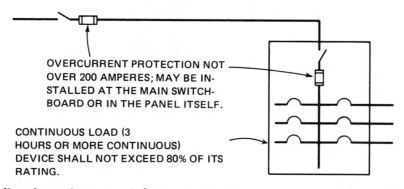

Fig. 20-8 When a panelboard contains snap switches rated at 30 amperes or less, the panelboard shall have overcurrent protection not to exceed 200 amperes. *NEC Section 384-16.*

If a 25% growth factor is added to the 94 amperes of actual loading, then the result is:

94 amperes X 125% = 118 amperes

Since the load consists of electric-discharge lighting, all four conductors must be considered to be current-carrying conductors, according to *NEC Table 310-16, Note 10.* Where four to six current-carrying conductors are installed in the same metal raceway, their maximum allowable load current must be reduced to 80%, as shown in the table to *Note 8, NEC Table 310-16.* Thus, in the bakery loading example, the derating requires a conductor with an ampacity of:

$$\frac{118 \text{ amperes}}{0.8} = 147 \text{ amperes}$$

According to *NEC Table 310-16* either a size 1, 90°C conductor (rated for 150 amperes) or a size 0, 75°C conductor (rated for 150 amperes) can be selected for the feeder.

Neutral Sizing (*Section 220-22*)

Under certain conditions, the neutral wire of a feeder may be smaller in size than the phase wires. This reduction in size is permissible because the neutral carries only the unbalanced portion of the load. However, on a three-phase, four-wire system, the load value is a vector sum (see figure 20-14). The table in figure 20-10 shows several conditions that may

exist. Conditions I, II, III, and VII are commonly accepted, but conditions IV, V, and VI appear to be contradictory. To find the unbalanced neutral current in a three-phase, four-wire system, only the loads connected between the neutral and phase wires are considered. In addition, equal power factors are assumed. Therefore, the equation used to find the unbalanced neutral current is as follows:

$$N = \sqrt{A^2 + B^2 + C^2 - AB - AC - BC}$$

where N = current in the neutral
A = current in phase A
B = current in phase B
C = current in phase C

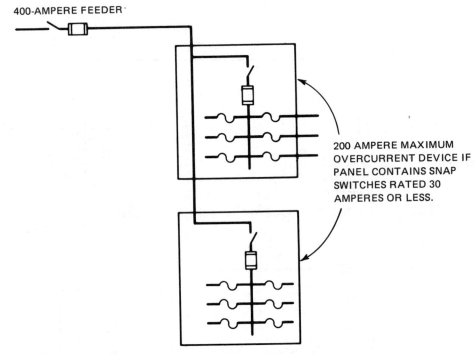

400-AMPERE FEEDER

200 AMPERE MAXIMUM OVERCURRENT DEVICE IF PANEL CONTAINS SNAP SWITCHES RATED 30 AMPERES OR LESS.

Fig. 20-9 Feed-through panel with adequate gutter space.

CONDITION	PHASE A	PHASE B	PHASE C	NEUTRAL
I	50	0	0	50
II	0	50	0	50
III	0	0	50	50
IV	50	50	0	50
V	50	0	50	50
VI	0	50	50	50
VII	50	50	50	0
VIII**	50	50	50	50 (approx.)

*Table is based on resistive loads.
**For inductive loads, such as fluorescent ballasts, the neutral conductor carries approximately the same value of current as the phase conductors. In this case, the neutral is considered to be a current-carrying conductor. See *Note 10* to *NEC Tables 310-16* through *310-19.*

Fig. 20-10 Load in amperes neutral unbalanced loading* three-phase, four-wire wye system.

To determine the maximum load on the neutral, the loads on each phase are tabulated (three-phase or single-phase, 208-volt loads are not included). Using the drugstore as an example, the panelboard schedule shows that the total volt-ampere load on each phase is:

Phase A	Phase B	Phase C
1,800	1,800	1,800
1,080	720	900
1,500	1,500	1,200
1,500	360	3,900
1,200	1,600	
7,080	5,980	

Since the least desirable situation is to have all of the loads on Phases B and C shut off, the load on Phase A is the value to be used for the sizing of the neutral. Therefore, for the drugstore (where Phases B and C are unloaded and all of the load on Phase A is energized), the neutral has a maximum load of:

7,080 volt-amperes ÷ 120 volts = 59 amperes
After a 25% growth factor is added, and the maxi-

mum load permitted by *Note 8* and *Note 10* of *NEC Table 310-16* is calculated because of the feeder having four current-carrying conductors in the raceway, the resulting load value is:

$$59 \text{ amperes} \times 1.25 = 73.75 \text{ amperes}$$

$$\frac{73.75}{0.8} = 92 \text{ amperes}$$

According to *NEC Table 310-16* either a No. 4, 90°C conductor (rated for 95 amperes) or a No. 3, 75°C conductor (rated for 100 amperes) can be selected as the neutral conductor. When the neutral conductor size is compared to the No. 1 size conductor required for the phase conductor, it can be seen that the neutral size reduction represents an economic advantage. In other words, the reduction in size of one conductor often means that a reduction can be made in the conduit size. However, the neutral conductor cannot be reduced in size in situations where two phases and the neutral of a three-phase, four-wire system are used to serve a single-phase panelboard. An example of this type of installation is

DOCTOR'S OFFICE LOADING SCHEDULE

Loading	NEC Article	NEC Minimum	Actual Allowance	Design Value
GENERAL LIGHTING				
Office Area:				
456.75 sq. ft. × 3.5 watts/sq. ft.	*220-2(b)*	1,599		
4 Style G luminaires			400	
6 Style H luminaires			600	
4 Style F luminaires			800	
Total general lighting		1,599	1,800	
Value to be used				1,800 VA
MOTOR LOAD				
16.8 amperes × 208 volts		3,494	3,494	
3.7 amperes × 208 volts		770		
25% allowance of larger motor	*440-32*	874		
Total motor load		5,138	4,264	
Value to be used				5,138 VA
OTHER				
Receptacle outlets				
10 outlets × 180 volt-ampere/outlet	*220-2(b)*	1,800		
Special outlet (sterilizer) maximum load	*220-2(b)*	3,000	3,000	
Total other load		4,800	3,000	
Value to be used				4,800 VA
Total for area				11,738 VA

Fig. 20-11

the doctor's office in the commercial building. The loading schedule for the doctor's office, figure 20-11, indicates a total load of 11,738 volt-amperes. With the addition of a growth factor, this load can be served by No. 4 feeder conductors. The neutral conductor size cannot be reduced because this conductor may, at times, carry both the unbalance of the doctor's office panelboard and an unbalance from the phase which is not connected to this particular panelboard.

Figures 20-12, 20-13, and 20-14 illustrate examples of the use of vectors to determine the approximate current in the neutrals of a three-phase, four-wire system.

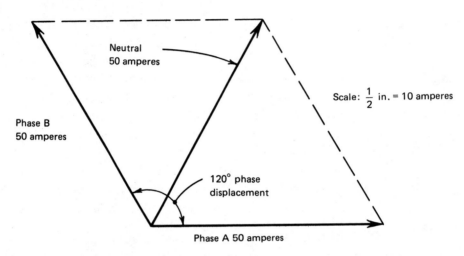

Fig. 20-12 Vector sum as shown in figure 20-10, Condition IV; loads are connected line-to-neutral.

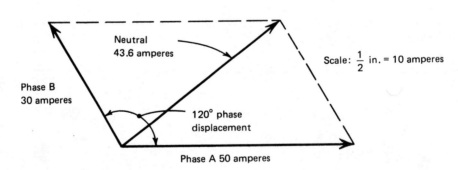

Fig. 20-13 Vector sum showing current in neutral when Phase A is carrying 50 amperes and Phase B is carrying 30 amperes; neutral is carrying approximately 43.6 amperes; loads are connected line-to-neutral.

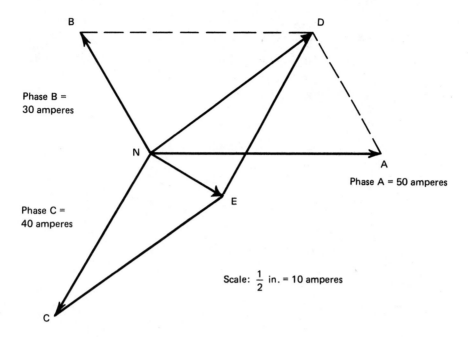

Phase B =
30 amperes

Phase A = 50 amperes

Phase C =
40 amperes

Scale: $\frac{1}{2}$ in. = 10 amperes

1. Draw each of the three-phase currents to scale, 120° apart.

2. Draw a parallelogram using any two phases, for example, NBDA.

3. Draw a diagonal for the parallelogram in Step 2: the diagonal is ND.

4. Draw a parallelogram using sides NC and ND; the result is parallelogram NDEC.

5. Draw a diagonal for parallelogram NDEC; the diagonal is NE.

6. Measure the length of NE. This length represents the approximate current in the neutral conductor for the three-phase, four-wire wye system. NE is approximately equal to 17 amperes. This example is for resistive loads only.

Fig. 20-14 Drawing vectors to determine approximate current in neutral of a three-phase, four-wire wye system; loads are connected line-to-neutral.

──REVIEW──────────────────────────

Note: Refer to the *National Electrical Code* or the plans as necessary.

1. To qualify as a lighting and appliance branch-circuit panelboard, a 42-circuit panelboard must have at least _____ overcurrent devices of 30 amperes or less.

2. A 225-ampere lighting and appliance branch-circuit panelboard with a connected load of 150 amperes must be protected at not more than _____ amperes.

3. A neutral of a three-phase, four-wire system will carry a current of _____ amperes when the currents in the phases are as follows: phase A, 12 amperes; phase B, 4 amperes; and phase C, 2 amperes. It is assumed that the loads are all connected neutral-to-phase.

4. The preferred colors to be used on a four-wire circuit are _____ , _____ , _____ , and _____ .

5. Determine the neutral size for the beauty salon, in amperes. Show all calculations.

(STUDENT CALCULATIONS)

APPENDIX

ELECTRICAL SPECIFICATIONS

General Clauses and Conditions

The General Clauses and Conditions and the Supplementary General Conditions are hereby made a part of the Electrical Specifications.

Scope

The electrical contractor shall furnish all labor and material to install the electrical equipment shown on the drawings herein specified, or both. The contractor shall secure all necessary permits and licenses required and in accepting the contract agrees to have all equipment in working order at the completion of the project.

Materials

All materials used shall be new and shall bear the label of the Underwriters Laboratories, Inc. and shall meet the requirements of the drawings and specifications.

Workmanship

All electrical work shall be in accordance with the requirements of the *National Electrical Code,* shall be executed in a workmanlike manner, and shall present a neat and symmetrical appearance when completed.

Show Drawings

The contractor shall submit for approval descriptive literature for all equipment installed as part of this contract.

Motors

All motors shall be installed by the contractor furnishing the motor. All wiring to the motors and the connection of motors will be completed by the electrical contractor. All control wiring will be the responsibility of the contractor furnishing the equipment to be controlled.

Wiring

The wiring shall meet the following requirements.

a. The minimum size conductors will be No. 12 AWG.
b. All conductors will be installed in conduit.
c. Three spare 1/2-inch conduits shall be installed from each flush-mounted panel into the ceiling joist space above the panel.
d. Branch-circuit conduits shall be electrical metallic tubing.
e. Feeder conduits shall be rigid metal conduit.
f. Receptacles shall be a grounding type and will comply with NEMA standards.

Switches

Switches shall be ac general-use, snap switches, specification grade, 20 amperes, 120-277 volts, either single-pole, three-way or four-way, as shown on the plans.

Time Clock

A time clock shall be installed to control the lighting in the front and rear entries. The clock will be connected to Panel EM circuit No. 1. The clock will be 120 volts, one circuit, with astronomic control and a spring-wound carry-over mechanism.

Receptacles

The electrical contractor shall furnish and install, as indicated in figure S1 (Electrical Symbol Schedule), specification-grade receptacles meeting NEMA standards and listed by Underwriters Laboratories, Inc.

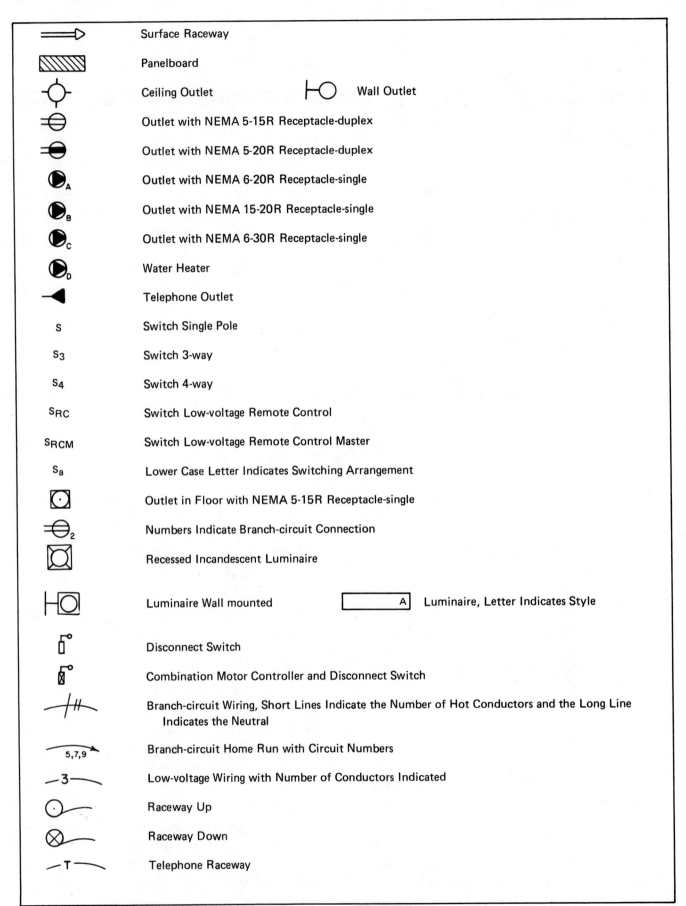

Fig. S1 Electrical symbols

Panelboards

The electrical contractor shall furnish panelboards, as shown on the plans and detailed in the panelboard schedules following these Specifications. Panelboards shall be listed by the Underwriters Laboratories, Inc. All interiors will have 225-ampere bus with lugs for incoming conductors. Boxes will be of galvanized sheet steel and will provide wiring gutters, as required by the *National Electrical Code*. Fronts will be suitable for either flush or surface installation and shall be equipped with a keyed lock and a directory card holder.

Molded Case Circuit Breakers

Molded case circuit breakers shall be installed in branch-circuit panelboards as indicated on the panelboard schedules following the Specifications. Each breaker shall provide inverse time delay under overload conditions and magnetic tripping for short circuits. The breaker operating mechanism shall be trip-free and multipole units will open all poles if one pole trips. Breakers shall have sufficient interrupting capacity to interrupt 10,000 rms symmetrical amperes.

Motor-generator

The electrical contractor will furnish and install a motor-generator plant capable of delivering 12 kVA at 120/208 volts, three phase. Motor shall be for use with gasoline or low-pressure gas, liquid cooled, complete with 12-volt batteries and battery charger, mounted on antivibration mounts with all necessary accessories, including mufflers, exhaust piping, fuel tanks, fuel lines, and remote derangement annunciator.

Transfer Switch

The electrical contractor will furnish and install a complete automatic load transfer switch capable of handling 12 kVA at 120/208 volts, three phase. The switch control will sense a loss of power on any phase and signal the motor-generator to start. When emergency power reaches 80% of voltage and frequency, the switch will automatically transfer to the generator source. When the normal power has been restored for a minimum of five minutes, the switch will reconnect the load to the regular power and shut off the motor-generator. Switch shall be in a NEMA 1 enclosure.

Luminaires

The electrical contractor will furnish and install luminaires, as described in the schedule shown in figure S2. Luminaires shall be complete with diffusers and lamps where indicated. Ballast, when required, shall be Class P and have a sound rating of A. Each ballast is to be individually fused; fuse size and type as recommended by the ballast manufacturer.

Service Entrance and Equipment

The electrical contractor will furnish and install a service entrance as shown on the Electrical Power Distribution Riser Diagram, figure S3. Transformers and all primary service will be installed by the power company. The electrical contractor will provide a concrete pad and conduit rough-in as required by the power company.

The electrical contractor will install a ground field adjacent to the transformer pad consisting of three 8-feet long copper weld rods connected to the grounded conductor. In addition, a connection shall be made to the main water pipe.

The electrical contractor will furnish and install a service lateral consisting of three sets of three 500 MCM and one 3/0 XHHW conductors, each set in a three-inch rigid metallic conduit.

The electrical contractor will furnish and install service-entrance equipment, as shown in figure S3 and detailed herein. The equipment will consist of nine switches, and the metering equipment will be fabricated in three type NEMA 1 sections. A continuous neutral bus will be furnished for the length of the equipment and shall be isolated except for the main bonding jumper to the grounding bus which shall also be connected to each section of the service-entrance equipment and to the water main. The switchboard shall be braced for 50,000 amperes rms symmetrical.

The metering will be located in one section and shall consist of seven meters. Five of these meters shall be for the occupants of the building and two meters shall serve the owner's equipment.

Style	Size	Lamps	Diffuser	Mounting	Description
A	4' x 18"	4-F40CW	Acrylic wrap-around	Surface	Shallow construction
B	2' x 2'	4-F20T12/CW	Opal Glass	Surface	Solid sides
C	50 9/10" x 8 5/8"	2-F40CW	Acrylic, clear, gasketed	Surface	Totally enclosed, contamination-proof
D	8'	2-F40CW	1/2" x 1/2" cell	Surface	2' x 2' luminous ceiling
E1	4' x 4 1/4"	8-50R20	None	Surface	Linear 6" deep
E2	4' x 4 1/4"	1-F40CW	None	Surface	Linear 6" deep
E3	4 1/2" x 4 1/4"	1-75R30	None	Stem	Square, matching
F	2' x 4'	4-F40CW	Glass, prismatic	Recessed	For plaster and tile ceiling
G	2' x 2'	2-F40CW/U	Glass, prismatic	Surface	Same as F
H	1' x 4'	2-F40CW	Glass, prismatic	Surface	Same as F
I	8'	2-F40CW	3/8 cube	Surface	Corridor use
J	12" sq.	1-150/99CL	Convex glass	Recessed	Vertical lamp, 6" deep
K		1-H37-5KL/DX	Prismatic	Vertical Surface	Aluminum reflector with photocell
L	4'	2-F40CW	None	Chain	Industrial, white enamel, all steel
M1		1-150R/SP	None	Surface	Adjustable canopy
M2		2-150R/SP	None	Surface	Matching M1

Fig. S2 Luminaire schedule

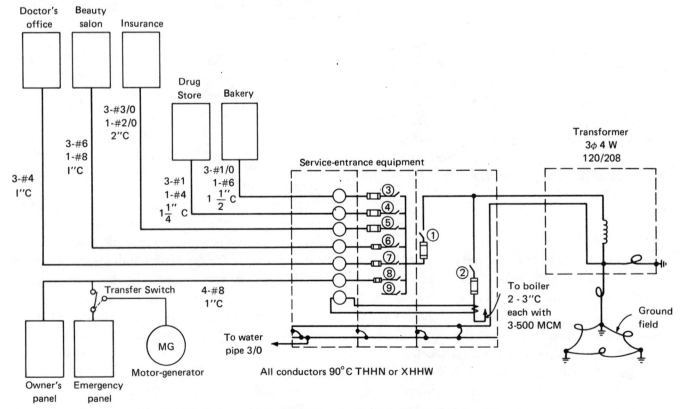

Fig. S3 Commercial building electrical power distribution riser diagram.

The switches shall be as follows:

1. Bolted pressure switch, three-pole, 600 amperes with three 600-ampere fuses.
2. Bolted pressure switch, three-pole, 800 amperes with three 800-ampere fuses.
3. Quick make-quick break switch, three-pole, 200 amperes, with three 175-ampere fuses.
4. Quick make-quick break switch, three-pole, 200 amperes with three 150-ampere fuses.
5. Quick make-quick break switch, three-pole, 200 amperes with three 200-ampere fuses.
6.,7. Quick make-quick break switch, a twin three-pole, 100-ampere switch with three 70- and two 90-ampere fuses.
8.,9. Quick make-quick break switch, a twin three-pole, 100-ampere switch with three 50-ampere fuses (switch 9 is a spare).

In addition, space will be available for a three-pole, 100-ampere quick make-quick break switch.

The bolted pressure switches shall be knife-type switches constructed with a mechanism which automatically applies a high pressure to the blade when the switch is closed. Switch shall be rated to interrupt 200,000 symmetrical rms amperes when used with current-limiting fuses having an equal rating.

The quick make-quick break switches shall be constructed with a device which assists the operator in opening or closing the switch to minimize arcing. Switches shall be rated to interrupt 200,000 symmetrical rms amperes when used with fuses having an equal rating. They shall have rejection type R fuse clips for Class R fuses.

Fuses

a. Fuses 601 amperes and larger shall have an interrupting rating of 200,000 amperes rms symmetrical. They shall provide time delay of not less than 4 seconds at 500% of the ampere rating. They shall be current limiting and of silver-sand Class L construction.

b. Fuses 600 amperes and less shall have an interrupting rating of 200,000 amperes rms symmetrical. They shall be of the rejection Class RK1 type. When protecting individual motor circuits and feeders to panels serving a combination of lighting and motor loads, these fuses shall be of dual-element construc-

tion having a time delay of not less than 10 seconds at 500% of the ampere rating. When protecting lighting panels, individual circuit breakers, and resistance heating elements, they shall have the minimum time delay.

c. Plug fuses shall be dual-element, time-delay type with Type S base.

d. All fuses shall be so selected to assure positive selective coordination as indicated on the schedule.

e. All lighting ballasts (fluorescent, mercury vapor, or others) shall be protected on the supply side with an appropriate fuse in a suitable approved fuseholder mounted within or on the fixture. Fuses shall be sized as recommended by the ballast manufacturer.

f. Spare fuses shall be provided in the amount of 20% of each size and type installed, but in no case shall less than three spares of a specific size and type be supplied. These spare fuses shall be delivered to the owner at the time of acceptance of the project, and shall be placed in a spare fuse cabinet mounted near the main switchboard.

g. Fuse identification labels, showing the size and type of fuses installed, shall be placed inside the cover of each switch.

Low-voltage Remote-control Switching

A low-voltage, remote-control switch system shall be installed in the drugstore as shown on the plans and detailed herein. All components shall be specification grade and constructed to operate on 24-volt control power. The transformer shall be a 120/24-volt, energy-limiting type for use on a Class II signal system.

Cabinet. A metal cabinet matching the panelboard cabinets shall be installed for the installation of relays and other components. A barrier will separate the control section from the power wiring.

Relay. A 24-volt ac, split-coil design relay rated to control 20 amperes of tungsten or fluorescent lamp loads shall be provided.

Switches. The switches shall be complete with wall plate and mounting bracket. They shall be normally open, single-pole, double-throw, momentary contact switches with on-off identification.

Rectifiers. A heavy-duty silicon rectifier with 7.5-ampere continuous duty rating shall be provided.

Wire. The wiring shall be in two- or three-conductor, color coded, No. 20 AWG wire.

Rubber grommet. An adapter will be installed on all relays to isolate the relay from the metal cabinet to reduce noise.

Switching schedule. Connections will be made to accomplish the lighting control as shown in the switching schedule, figure S4.

HEATING AND AIR-CONDITIONING SPECIFICATIONS

Only those sections which pertain to the electrical work are listed here.

Motors

All motors will be installed by the contractor furnishing the motor. All electrical power wiring to the motors and the connection of the motors shall be made by the electrical contractor. All control wiring and the control devices will be the responsibility of the contractor furnishing the equipment to be controlled.

Boiler

The heating contractor will furnish and install a 225-kW electric hot water heating boiler completely equipped with safety, operating, and sequencing controls for 208-volt, three-phase electric power.

Hot Water Circulating Pumps

The heating contractor will furnish and install five circulating pumps. Each pump will serve a

Switch	Relay	Area Served	Branch Circuit #
RCa	A	Main area lighting	1
RCb	B	" " "	1
RCc	C	" " "	3
RCd	D	" " "	3
RCe	E	" " "	5
RCf	F	" " "	5
RCg	G	Makeup area	7
RCh	H	Storage	7
RCi	I	Toilet	7
	J	Pharmacy	9
	K	"	11
	L	"	13
RCm	M	Stairway	7
	N	Show window	15
	O	Sign	15
RCp	P	Basement area	7
RCM-1	G		
RCM-2	H, I		
RCM-3	J, K, L		
RCM-4	L, O, N		
RCM-5	A, B, C, D, E, F		
RCM-6	M, P		
RCM-7	N		
RCM-8	O		

Fig. S4 Drugstore low-voltage, remote-control switching schedule

separate rental area and be controlled from a thermostat located in that area as indicated on the electrical plans. Pumps will be 1/6 horsepower, 120 volts, single phase. Overload protection will be provided by a manual motor starter.

Air-conditioning Equipment

Air-conditioning equipment will be furnished and installed by the heating contractor in four of the rental areas indicated as follows:

Drugstore. A split system packaged unit will be installed with a rooftop compressor-condenser and a remote evaporator located in the basement. The electrical characteristics of this system are:

Voltage: 208 volts, three phase, three wire
Compressor: FLA 20.2; LRA 90.0
Condenser: FLA 3.2; 1/4 horsepower, single phase
Evaporator: FLA 3.2; 1/4 horsepower, single phase

Insurance Office. A single package unit will be installed on the roof with electrical data identical to the unit specified for the drugstore.

Beauty Salon. A single package unit will be installed on the roof and will have the following electrical characteristics:

Voltage: 208 volts, three phase, three wire
Compressor: FLA 14.1; LRA 61.5
Condenser-
  evaporator
  motor: FLA 3.3; 1/4 horsepower, single phase

Doctor's Office. A single package unit will be installed on the roof and will have the following electrical characteristics:

Voltage: 208 volts, single phase
Compressor: FLA 16.8; LRA 60
Condenser-
  evaporator
  motor: FLA 3.7; 1/3 horsepower, single phase

Heating Control

In cooled areas, the heating and cooling will be controlled by a combination thermostat located in the proper area as shown on the electrical plans.

PLUMBING SPECIFICATIONS

Only those sections which pertain to the electrical work are listed here.

Motors

All motors will be installed by the contractor furnishing the motor. All electrical power wiring to the motors and the connection of the motors shall be made by the electrical contractor. All control wiring and control devices will be the responsibility of the contractor furnishing the equipment to be controlled.

Sump Pump

The plumbing contractor will furnish and install an electric motor-driven, fully automatic sump pump. Motor will be 1/2 horsepower, 208 volts, single phase. Overload protection will be provided by a manual motor starter.

PANELBOARD SCHEDULES

WIRE COLOR	MIN. WIRE SIZE	CIRCUIT NO.	PROTECTION	PHASE	VA LOAD	M-MOTOR R-RECEPTACLE LOADING L-LIGHTING		PANEL POLES
B	12	1	20	A	1800	L	Sales Area North	1
B	12	2	20	A	1080	R	Sales Area and Basement North	2
R	12	3	20	B	1800	L	Sales Area Center	3
R	12	4	20	B	720	R	Toilet, Storage, Makeup Area — Basement West	4
BU	12	5	20	C	1800	L	Sales Area South	5
BU	12	6	20	C	900	R	Sales Area and Basement South	6
B	12	7	20	A	1500	L	Stair, Makeup, Storage, Toilet, Basement	7
B	12	8	20	A	1500	R	Show Window	8
R	12	9	20	B	1500	L	Pharmacy	9
R	12	10	20	B	360	R	Pharmacy	10
BU	12	11	20	C	1200	L	Pharmacy	11
	12	12	20	C		Spare		12
B	12	13	20	A	1200	L	Pharmacy	13
		14	20	A		Spare		14
R		15	20	B	1600	L	Show Window and Sign	15
B	8			B				16
R	8	16	40	C	8602	M	Cooling System 3 Pole	17
BU	8			A				18
		17	20	C		Spare		19
		18	20	B		Spare		20
		19	20	A		Spare		21
		20		C		Space		22
		21	20	B		Spare		23
		23		C		Space		24
								25

Panel legend:
BU–BLUE
B–BLACK
R–RED

Drugstore PANEL
120/208 VOLTS 3 PH 4 W

WIRE COLOR	MIN. WIRE SIZE	CIRCUIT NO.	PROTECTION	PHASE	VA LOAD	LOADING	PANEL POLES
						Bakery PANEL — 120/208 VOLTS 3 PH 4 W	
B		1	20	A	1800	L Bake Area	1
B		2	20	A	900	R Southwall	2
R		3	20	B	960	L Sales Area	3
R		4	20	B	1398	R Northwall L – Toilet M – Exhaust Fan	4
BU		5	20	C	1030	L Show Window – Sign	5
BU		6	20	C	1350	L Basement	6
B				A			7
R		7	20	B	4914	M Cake Mixers and Dough Divider 3 Pole	8
BU				C			9
B		8	20	A	1500	R Show Window	10
B				A			11
R		9	20	B	2792	M Doughnut Machine 3 Pole	12
BU				C			13
R		10		B	380	L Night Lights	14
		11		A		Spare	15
		12		C		Spare	16
		13		B		Spare	17
B				A			18
R		14	60	B	16000	M Oven 3 Pole	19
BU				C			20
		15		C		Space	21
		16		A		Space	22
		18		B		Space	23
		20		C		Space	24
							25

WIRE COLOR	MIN. WIRE SIZE	CIRCUIT NO.	PROTECTION	PHASE	VA LOAD	LOADING	
						Owner's PANEL — 208 VOLTS 3 PH 4 W	
		1	20	A		Spare	1
		2	20	A		Spare	2
		3	20	B		Spare	3
		4	20	B		Spare	4
BU	12	5	20	C	945	L Exterior	5
BU	12	6	20	C	1250	L Corridor, Toilets, Janitor Closet	6
		7		A		Space	7
		8		A		Space	8
		9		B		Space	9
		10		B		Space	10
							11

BU–BLUE B–BLACK R–RED

M–MOTOR R–RECEPTACLE L–LIGHTING

Beauty Salon **PANEL**

120/208 **VOLTS** *3* **PH** *4* **W**

BU-BLUE
B-BLACK
R-RED

M-MOTOR
R-RECEPTACLE
L-LIGHTING

WIRE COLOR	MIN. WIRE SIZE	CIRCUIT NO.	PROTECTION	PHASE	VA LOAD	LOADING	PANEL POLES
B	12	1	20	A	1080	R General	1
B	12	2	20	A	1300	L General	2
R	12	3	20	B	1200	R Hair Dryers	3
R	12	4	20	B	900	L Chairs	4
BU	10			C			5
B	10	5	30	A	5742	M Cooling System 3 Pole	6
R	10			B			7
		6	20	C		Spare	8
		7	20	C		Spare	9
		8	20	B		Spare	10
		9	20	A		Spare	11
R	10	10	30	B	4000	R Water Heater 2 Pole	12
BU	10			C			13
		11		B		Space	14
		12		A		Space	15
		13		C		Space	16
		14		B		Space	17
		16		C		Space	18
							19

Doctor's Office **PANEL**

120/208 **VOLTS** *1* **PH** *3* **W**

BU-BLUE
B-BLACK
R-RED

M-MOTOR
R-RECEPTACLE
L-LIGHTING

WIRE COLOR	MIN. WIRE SIZE	CIRCUIT NO.	PROTECTION	PHASE	VA LOAD	LOADING	PANEL POLES
B	12	1	20	B	400	L Waiting Room	1
B	12	2	20	B	720	R Waiting Room	2
R	12	3	20	C	1400	L Examining Room	3
R	12	4	20	C	1080	R Examining Room	4
B	10	5	30	B	3000	R Sterilizer 2 Pole	5
R	10			C			6
B	10	6	40	B	4264	M Cooling System 2 Pole	7
R	10			C			8
		7	20			Spare	9
		8	20			Spare	10
		9	20			Spare	11
		10	20			Spare	12
		11	20			Spare	13
		12	20			Spare	14
							15

BU–BLUE
B–BLACK
R–RED

Insurance Office PANEL
120/208 VOLTS 3 PH 4 W

WIRE COLOR	MIN. WIRE SIZE	CIRCUIT NO.	PROTECTION	PHASE	VA LOAD	LOADING (M-MOTOR / R-RECEPTACLE / L-LIGHTING)	PANEL POLES
B	12	1	20	A	1600	L Staff Office	1
B	12	2	15	A	900	R Clerical Office	2
R	12	3	20	B	1400	L Staff Office	3
R	12	4	15	B	900	R Clerical Office	4
BU	12	5	20	C	1000	L Staff Office	5
BU	12	6	15	C	900	R Clerical Office	6
B	12	7	20	A	540	R Staff Office Floor	7
B	12	8	15	A	900	R Staff Office	8
R	12	9	20	B	1800	L Clerical Office	9
R	12	10	15	B	900	R Staff Office	10
BU	12	11	20	C	1000	L Clerical Office	11
BU	12	12	15	C	900	R Staff Office	12
B	8			A			13
R	8	13	40	B	8602	M Cooling System 3 Pole	14
BU	8			C			15
B	12	14	20	A	540	R Staff Office	16
B	10	15	30	A	4992	M Copy Machine 2 Pole	17
R	10			B			18
R	12	16	20	B	900	R Staff Office	19
		17	20	C		Spare	20
BU	12	18	20	C	900	R Staff Office	21
		19	20	A		Spare	22
B	12	20	20	A	720	R Staff Office	23
		21	20	B		Spare	24
B	12	22	20	B	1080	R Staff Office	25
		23	20	C		Spare	26
BU	12	24	20	C	1080	R Client Area	27
		26	20	A		Spare	28
		28	20	B		Spare	29
		30	20	C		Spare	30
				A		Space	31
				B		Space	32
				C		Space	33
				A		Space	34
				B		Space	35
				C		Space	36
							37

WIRE COLOR	MIN. WIRE SIZE	CIRCUIT NO.	PROTECTION	PHASE	VA LOAD	Emergency PANEL 120/208 VOLTS 3 PH 4 W		PANEL POLES
						M-MOTOR R-RECEPTACLE LOADING L-LIGHTING		
B	12	1	20	A	600	L	Entry Front and Back	1
B	12	2	20	A	600	L	Utility Area Basement	2
R	12	3	20	B	1250	R	Telephone	3
R	12	4	20	B	720	R	Utility Area	4
BU	12	5	20	C	1250	R	Telephone	5
	12	6	15	C			Spare	6
B	12	7	15	A	528	M	Water Circulating Pump Drugstore	7
B	12	8	15	A	528	M	Water Circulating Pump Bakery	8
R	12	9	15	B	528	M	Water Circulating Pump Insurance	9
R	12	10	15	B	528	M	Water Circulating Pump Beauty Salon	10
BU	12	11	15	C	528	M	Water Circulating Pump Doctor's Office	11
BU	12	12	20	C	1123	M	Sump Pump	12
	12	13	20	A			Spare	13
	12	14	20	A			Spare	14
	12	15	20	B			Spare	15
		16	20	B			Spare	16
		17		C			Space	17
		18		C			Space	18

BU-BLUE
B-BLACK
R-RED

USEFUL FORMULAS

TO FIND	SINGLE PHASE	THREE PHASE	DIRECT CURRENT
AMPERES when kVA is known	$\dfrac{kVA \times 1000}{E}$	$\dfrac{kVA \times 1000}{E \times 1.73}$	not applicable
AMPERES when horsepower is known	$\dfrac{hp \times 746}{E \times \% \text{ eff.} \times pf}$	$\dfrac{hp \times 746}{E \times 1.73 \times \% \text{ eff.} \times pf}$	$\dfrac{hp \times 746}{E \times \% \text{ eff.}}$
AMPERES when kilowatts are known	$\dfrac{kW \times 1000}{E \times pf}$	$\dfrac{kW \times 1000}{E \times 1.73 \times pf}$	$\dfrac{kW \times 1000}{E}$
KILOWATTS	$\dfrac{I \times E \times pf}{1000}$	$\dfrac{I \times E \times 1.73 \times pf}{1000}$	$\dfrac{I \times E}{1000}$
KILOVOLT-AMPERES	$\dfrac{I \times E}{1000}$	$\dfrac{I \times E \times 1.73}{1000}$	not applicable
HORSEPOWER	$\dfrac{I \times E \times \% \text{ eff.} \times pf}{746}$	$\dfrac{I \times E \times 1.73 \times \% \text{ eff.} \times pf}{746}$	$\dfrac{I \times E \times \% \text{ eff.}}{746}$
WATTS	$E \times I \times pf$	$E \times I \times 1.73 \times pf$	$E \times I$

I = amperes   E = volts   kW = kilowatts   kVA = kilovolt-amperes

hp = horsepower   % eff. = percent efficiency   pf = power factor

EQUATIONS BASED ON OHM'S LAW

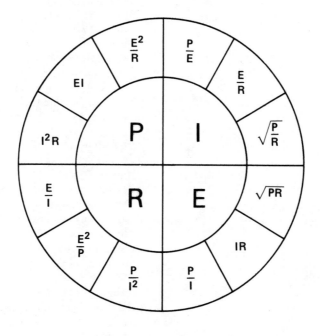

P = POWER IN WATTS

I = CURRENT IN AMPERES

R = RESISTANCE IN OHMS

E = ELECTROMOTIVE FORCE IN VOLTS

INDEX